Library of
Davidson College

The Nitrate Industry and Chile's Crucial Transition: 1870-1891

THE NITRATE INDUSTRY AND CHILE'S CRUCIAL TRANSITION: 1870-1891

Thomas F. O'Brien

New York University Press
New York *and* London
1982

Copyright © 1982 by New York University

Library of Congress Cataloging in Publication Data

O'Brien, Thomas F., 1947–
The nitrate industry and Chile's crucial transition, 1870–1891.

Bibliography: p.
Includes index.
1. Nitrate industry–Chile–History–19th century. 2. Chile–Economic conditions. I. Title.
HD9660.N52C56   338.2'764'0983   82-2131
ISBN 0-8147-6159-3   AACR2

Clothbound editions of New York University Press books
are Smyth-sewn and printed on permanent and durable acid-free paper.

Manufactured in the United States of America

To Diane

Contents

Map: The Nitrate Zone   6
Tables   ix
Preface   xi
Acknowledgments   xv
I. Merchant Capital in the Nitrate Industry   1
II. The Peruvian Expropriation   26
III. Elites and Foreign Investors: Chilean Nitrate Policy   42
IV. Expansion and Foreign Penetration   63
V. The Symbiotic Relationship   77
VI. Men Making Their Own History   96
VII. The Emergence of Monopoly Capitalism   111
VIII. The State in Transition   124
IX. Conclusion   147
Appendix   157
Notes   163
Sources   193
Index   205

Tables

1. Nitrate Exports from Peru and Average Prices in England, 1861–79
2. Comparative Position of Chilean and European Nitrate Companies in Tarapacá, 1873
3. *Oficinas* Operating under Contract and Held by Original Owners, December 1878
4. Chilean Nitrate Exports (in Tons), 1880–86
5. Nitrate Production in Tarapacá, 1886
6. Share Prices of Leading Companies on the Chilean Stock Exchange, August 1877–86
7. Dividends of Leading Companies on the Chilean Stock Exchange, 1875–77, 1879–81, 1884–86
8. Participation of Ten Chilean Entrepreneurs in Chilean Nitrate Companies
9. The Estate of Antonio María Costa, 1888
10. The Estate of José María Necochea, 1890
11. The World Nitrate Market (in Tons), 1886-90
12. Share Prices of North's Nitrate Companies, 1889
13. Nitrate Production Companies Registered with the London Stock Exchange as of 12 December 1889
14. Directors of Nitrate Production Companies Serving on the Board of More Than One Company, 1888-89
15. State Revenues and Private Domestic Revenues (Current Pesos), 1881-90
16. Percentage of Nitrate Industry Controlled by Chilean and European Investors

Preface

The nitrate industry launched Chile into a prolonged era of prosperity in the last quarter of the nineteenth century. Yet like so many Third World countries of today, Chile's development eventually stalled in a process of dynamic stagnation—dynamic economic growth encumbered by precapitalist social structures and political instability. It became a society unable to achieve the self-generating social and economic changes so often identified with the concept of development. That transition into a condition of stalled development or peripheral capitalism can be traced to Chile's absorption of the nitrate industry.

In the first half century of independence, Chile proudly distinguished itself from its troubled neighbors by social and political stability as well as economic prosperity. It was not surprising, then, that Chileans played a major role in the nitrate industry of Peru and Bolivia, and finally seized the nitrate regions during the War of the Pacific (1879-83). Yet Chile's triumph soon proved to be a tarnished one. The nation entered a period of unprecedented prosperity, but Europeans monopolized the nitrate industry, and government stability collapsed in a bitter civil war. Decades of political bickering ensued, social conflict accelerated, and the promise of developmental industrialization remained unfulfilled.

Numerous scholars have attempted to explain the causes of Chile's frustrated development. With the foreign nitrate monopoly still emerging, the Peruvian Guillermo Billinghurst first pointed an accusing finger at the British "Nitrate King," John Thomas North, and his associates. Focusing on the Civil War of 1891, the Chilean Hernán Ramírez Necochea has expanded that attack, condemning North as the agent of British imperialism and adding regressive elements of the Chilean elite to the list of his accomplices. In rebuttal to Ramírez, the British historian Harold Blakemore has persuasively argued that the Civil War was an internal political dispute in which foreign nitrate interests were a convenient issue rather than a causal factor. Francisco Encina, Chile's own version of Count Gobineau, attributed the loss of the industry and subsequent problems to racial traits that made Chileans unfit for business enterprises. Based on a structuralist perspective, the economist Aníbal Pinto reached a similar conclusion, arguing that Chile's export economy created an elite incapable of the progressive activities required for domestic development of the industry. While Billinghurst focused on developments within the industry, these other authors have dealt with the industry as one element in a broader analysis of Chilean economic or political history. In terms of analytical perspectives, their works tend to fall into two broad categories—those that focus on internal factors and those that emphasize the importance of external agents.

This book offers a new perspective on these events, which marked a watershed in Chilean history. Based on new archival evidence, it details developments in the industry while it traces the causes and effects of the foreign nitrate monopoly to Chile's social and economic structures and its place in the world economy. Its analytical focus is the interaction between a traditional society and the capitalist center, a relationship that eventually transcended mere commercial interchange to ground itself in Chile's partial absorption of the capitalist mode of production. In this way the nitrate industry serves as a prism, drawing on the historical evolution of Chile to illuminate its development while shedding its own light upon Chilean history. If at all successful, this book will serve, as its predecessors have, as the starting point for further investigation and revision of its hypotheses.

Influenced, as were earlier writers, by Billinghurst's evidence, I originally assumed that the period prior to the War of the Pacific was no more than a preface to the crucial developments of the 1880s. However, my own research has convinced me that events of the 1870s had a decisive impact on the industry and Chile's relationship to it. As a result, the first three chapters deal with the initial thrust of Chilean and European capitalists into the industry in that period and with Chilean acceptance of European control during the War of the Pacific. The following four chapters detail the emergence of the foreign nitrate monopoly and the accommodation it achieved with Chile's traditional society after the war. Chapter 8 examines the Civil War of 1891 as a political adjustment necessitated by Chile's new relationship to the nitrate industry in particular and to European capitalism in general.

The reader may note that although the concept of dependency emerges as a central theme of this book, the term "dependency" rarely appears except in the "Conclusion." This is not the work of some sharp-eyed editor eager to banish redundancies. It is, rather, an expression of a growing concern with the concept of dependency and its application to Latin American history. The past two decades have produced an exciting and stimulating array of works by proponents and critics of the dependency hypothesis. The analysis in this book owes much to that lively debate, especially the contributions of Fernando H. Cardoso and Robert Brenner. Cardoso has traced the causes of underdevelopment to the interplay of internal social and political forces with the international activities of the advanced capitalist economies. Brenner has reemphasized the role of class conflict in the process of underdevelopment. But it is equally true that most of the debate continues to be conducted at the theoretical level. If the dependency thesis is ever to become an effective working hypothesis that can aid our investigation of Latin American history, it must be tested and refined through specific historical research. While readily admitting my own intellectual predisposition, I have tried to avoid substituting the term "dependency" as a simple catch-all explanation of the phenomena examined in this work. Hopefully, my effort to attain specificity in treating the causes and effects of foreign domination of the nitrate industry will serve as one small step in clarifying the dependency hypothesis.

Finally, this work is inspired by a concern that transcends the abstractions of historical and developmental analysis that are essential if sometimes tedious parts of its makeup. While conducting my research in Chile, I became caught up in the excitement of the flawed experiment in social justice that was the Popular Unity government. This book does not attempt to explain the failure of that noble effort or its tragic aftermath. Yet I hope it will offer some insight into the forces shaping the history of a people who have asked so little and endured so much.

Acknowledgements

In the preparation of this work I have been assisted by the talents and resources of a number of people and institutions. My deepest debt of gratitude is to my wife, Diane, for her faith and unfailing inspiration. Hugh Hamill created the open intellectual atmosphere that spawned the idea for this work and offered the encouragement that brought it to fruition. My friends and colleagues John Hart and Allen Woll made incisive critiques of the manuscript that improved it immeasurably. A special thanks is due Harold Blakemore, who not only has aided me in my research but whose own work has contributed so much to our understanding of the nitrate industry and its significance for Chilean history. Joseph Criscenti first sparked my interest in Latin American history. Paul Goodwin read early drafts of the manuscript and suggested salutory revisions. The Chilean scholars Jacqueline Garreaud and Oscar Bermúdez Miral provided both research assistance and hospitality.

The American Social Science Research Council and the Danforth Foundation generously contributed financial assistance. The professional skills of the staffs of the Biblioteca Nacional and Archivo Nacional in Santiago, and the Guildhall Library and Public Record Office in London, were a great asset in my research. Christine Womack of the University of Houston skillfully typed the manuscript.

I also wish to thank the editors of the *Journal of Latin American Studies,* and the *Hispanic American Historical Review* for permission to use materials previously published in those journals.

I am indebted in a very special way to the González family who taught Diane and me so much about Chile and her people. Finally, this work owes much to my parents, whose generosity, patience, and understanding helped us through the difficult times.

[1]

Merchant Capital in the Nitrate Industry

AT the beginning of the 1870s, Chilean and European capitalists became locked in intense competition in Peru's thriving nitrate industry. That direct competition with Europeans in the area of foreign investment was even attempted testifies to Chile's remarkably coherent evolution in the first three quarters of the nineteenth century. With other Latin American nations suffering through interminable political crises and severe economic dislocations, Chile prospered under the rule of a series of constitutionally elected administrations. Between 1831 and 1871 four presidents succeeded each other in orderly fashion, and the value of the nation's exports more than quadrupled. The country's small geographic extent, easily defensible borders, and close-knit ruling elite provided essential preconditions for the emergence of a stable central government and a strong national economy. The autocratic state fostered by the Chilean statesman Diego Portales further contributed to Chile's emergence as a governable sovereign nation. But most important in determining Chilean stability and prosperity was the pattern of economic relationships.

By the end of the eighteenth century the great rural estates, the social and economic anchor of the oligarchy, formed a pattern of landownership that remained virtually unchanged during the nineteenth century. The insignificance of church property and the isolation of the Indian population in the southernmost reaches of the country precluded disruptive assaults on corporate landholdings that occurred in Peru and Mexico. A surplus rural population provided landowners with an easily accessible labor supply and spared the nation the violent effort to control a scarce labor force that characterized Colombia at this time. Chile's relationship to the world economy reinforced this pattern of stability.

Within the Spanish economic empire, Chile had served as an appendage to the mining enclaves of the Viceroyalty of Peru. Lacking easily exploitable precious metals, the colony's scanty export trade was largely confined to shipments of wheat and livestock to its northern neighbor. Independence in 1817 did not significantly alter the pattern or volume of the country's exports. Urban merchants and rural landowners thus shared a commonality of interests that was virtually unaffected by the collapse of Spain's imperial system. The civil strife that divided Argentina between coastal interests tied to an expanding trade with Europe and groups in the interior bent on preserving internal markets was inconceivable in the Chilean context. Chile's marginal links to the European market also precluded the glut of cheap foreign imports and the desperate search for viable export markets that jarred other Latin American economies in the postindependence period.[1] By mid-century, however, Chile's ties to the international trading system began to increase significantly.

Silver, copper, and coal discoveries encouraged the export of Chile's mineral wealth. The discoveries at Chañarcillo in northern Chile in 1832 sparked the growth of silver mining, which received renewed impetus from the brief Caracoles boom in the early 1870s. Similar if less spectacular developments occurred in copper mining. Between 1844 and 1876 annual copper production increased from 10,865 tons to 57,660 tons. Exploitation of coalfields in the south also began with the principal producer, the Compañía de Lota, increasing its annual output from 7,815 tons in 1852 to 186,944 tons in 1876.[2]

During the same period, Chile's agricultural sector took its first tenuous steps into the emerging world trade system. Gold discoveries in California and Australia created markets for Chilean wheat and flour between 1850 and 1858. By 1868 the burgeoning industrial centers of England had become a new and more important destination for Chilean grain exports. The average annual value of agricultural exports grew from 3,756,000 pesos in the 1850s to 13,241,000 pesos in the early 1870s.[3]

Expansion of exports prompted the establishment of financial institutions by merchants in the port of Valparaíso. The first bank appeared in 1849, and by 1878 the country enjoyed the services of eleven banks with a paid-in capital of 19,157,588 pesos.[4] This period also witnessed the development of joint-stock companies as an accepted form of economic enterprise. By 1870, fifty-one such corporations were in existence with a total nominal capital of 66,068,000 pesos.[5] Economic growth prompted by

Chile's absorption into the world economy as a supplier of primary products had significant effects within the domestic social structure.

Throughout the nineteenth century a small (no more than two hundred families at mid-century), close-knit landowning elite dominated Chilean society.[6] Yet these families demonstrated a willingness to absorb both foreign and domestic nouveau riche elements created by growth in the export economy. Immigrant families such as the Edwards and Ross clans acquired their fortunes in Chile's prosperous economy and gradually achieved places of pominence in the oligarchy. With mineral exports regularly producing two or three times the earnings of agricultural exports, mining provided the most ready access to the upper social strata. Merchants and bankers also made the ascent of the social pyramid. New wealth, adorned with the proper trappings of a good marriage and the all-important symbol of a landed estate, not infrequently made its appearance in the upper ranks of society. Moreover, landowning families were not averse to engaging in commercial undertakings or investing in mining enterprises. The national credit market further strengthened the complex network of familial ties and economic interests that bound the elite together. Merchants regularly extended credit to both miners and agriculturists, supplementing extensive private credit transactions among the elite. As Arnold Bauer has noted, "Often related by blood or social ties, the miners, bankers, landowners and merchants forged an increasingly homogeneous group as the century progressed."[7] The oligarchy's absorption of the newly rich and its own enterprising endeavors assured it of virtually complete control of the domestic economy. However, Chile relied heavily upon the imposing commercial and financial power of Great Britain to link its export sectors to the international system of trade.

Great Britain's importance to the Chilean economy extended far beyond its role as a market for agricultural exports. By 1875, Great Britain and its empire absorbed 60 percent of Chile's total exports and provided 40 percent of its imports. Furthermore, 60 percent of the leading merchants in Chile were foreigners; the most important group was comprised of British trading houses such as Antony Gibbs and Sons Ltd., and Williamson, Balfour and Company. British interests had also penetrated the Chilean merchant marine. In addition, loans negotiated in London helped finance the operations of the Chilean government.[8] Despite foreign domination of international commerce and finance, Chilean diplomatic relations were a model of nineteenth-century decorum.

The stability of the Chilean state's relationships with Great Britain and other European powers is explained only in part by the economic prosperity that allowed Chile to pay its international debts in a timely fashion. More important was the clear division of functions, with the elite controlling the means of production and foreign commercial houses dominating the international trading network. The state was thus reduced to a limited role of regulating commercial activity. Domestic control of the means of production also contributed to internal political stability.

Although the cohesion of the elite offers a partial explanation for the relative passivity of Chilean political life, the timing of Chile's entrance into the international trading network and the limited economic functions of the state were at least as important. Coming at a time when the appetite of the European economies for foodstuffs and raw materials began to grow prodigiously, Chile's emergence as a producer of primary goods assured prosperity to its main economic sectors. Since this economic activity was largely a matter of private initiative, the state's interference in economic affairs could be kept to a bare minimum. These conditions preempted intraelite struggles to control state resources, so common in other Latin American countries. Even disputes over national economic policies were infrequent subjects of bitter political debate. In the long run, the rapid growth of the export economy did trigger some subtle changes in this stable but not static system.

Economic expansion spurred government revenues to an annual growth rate of 6.4 percent between 1830 and 1880.[9] With increasing resources at its command, the state undertook a series of projects that directly benefited the domestic economy. Such activities became particularly noticeable during the presidency of Manuel Montt (1851–61), during which the rail network was expanded, roads and ports were improved, and the telegraphic network was extended.[10] The growth of state services led inevitably to an expansion of the civil bureaucracy, which numbered 2,525 by 1860. Salaries were not spectacular, but public employment constituted an important form of patronage and political control exercised by the president.[11] But the balance of economic power still lay in the private sector. Furthermore, the intertwining social and economic interests of the elite preempted any fundamental split along the lines of economic sectors or rural versus urban interests. Although factions representing regional or familial interests did exist, these factors precluded political divisions that might threaten the institutional integrity of the state.

Chile's secondary status within the Spanish economic empire, the cohesion of its oligarchy, and the maintenance of domestic control of natural resources ensured its smooth transition to the role of a primary products supplier on the periphery of the world economy. The one clear sign of instability came from the export sector itself. As dependence on exports increased, so did the economy's susceptibility to sudden changes in world market conditions. The first sign of the economy's fragility came between 1857 and 1861 when a panic in Europe and the United States sparked a depression in Chile.[12] The warning was soon forgotten as the economy rebounded from the crisis. In fact, by the mid-1860s the development of a domestically controlled export economy linked to the rapidly expanding capitalist economies of Europe permitted Chile to develop important ties with Peru's nitrate enclave. Chilean penetration of the nitrate industry was also conditioned by structural factors within the industry itself and the Peruvian economy.

During the nineteenth century, the world's only commercially exploitable nitrate resources were located on the Pacific coast of South America between 19° and 26° south latitude. The coast of the region presents a seemingly endless panorama of steep slopes that plunge suddenly to meet the sea. Only occasionally is the pattern broken by small human settlements such as Iquique that serve, however inadequately, as seaports. The cliffs of the shore climb into a low coastal mountain range, which in turn dips into a sandy desert plain. The waterless wastes of this desert *pampa* stretch fifty kilometers inland to meet the foothills of the Andes. Here, mountain streams provide oases of vegetation. The Loa river, the only stream that reaches the coast, divides the great desert valley into two parts, the Pampa del Tamarugal to the north and the Atacama Desert to the south (see map, p. 6). It was in the area north of the river, in the Peruvian province of Tarapacá, that the most important nitrate deposits were located.

The nitrate deposits lie approximately twenty kilometers inland at the foot of the coastal range. Nitrate of soda is most commonly encountered in layered formations containing varying grades of nitrate. The caliche or layer with the highest nitrate content lies anywhere from one to six meters below the desert floor. It ranges in consistency from brittle to a hardness that only high explosives can shatter.[13]

Between 1810 and 1812 seven or eight *oficinas* (nitrate refineries) employing the *parada* system were built in Tarapacá. In this manually operated system pulverized caliche was boiled in water dissolving the nitrate, or *salitre,* which then settled out as the liquid cooled. But the industry remained at a primitive level of development until a large market for its product developed in Europe.

The use of chemical fertilizers in European agriculture had increased sufficiently by the 1830s to make nitrates a useful return cargo for ships

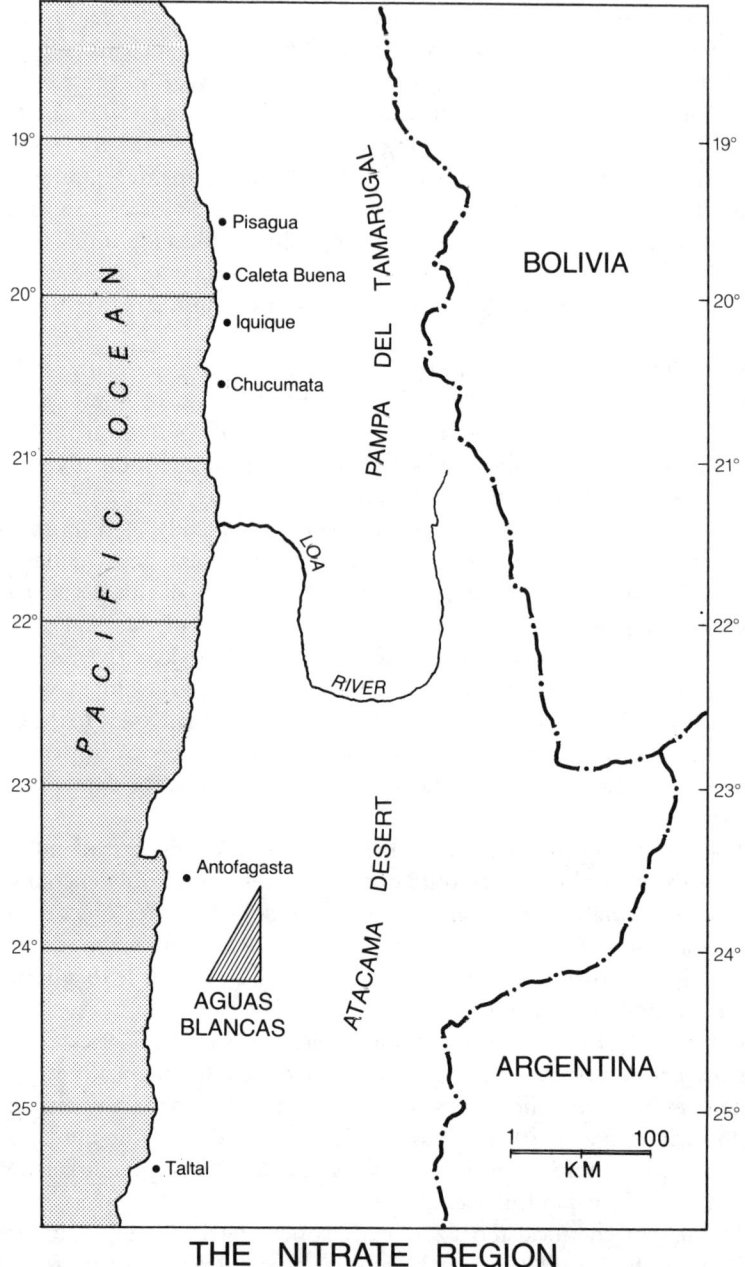

THE NITRATE REGION

Source: Oscar Bermudez Miral, *Historia del salitre desde sus orígenes hasta la Guerra del Pacífico* (Santiago, 1963), p. 16.

sailing to Europe. Even with the development of more easily extracted guano deposits on Peru's Chincha Islands, the growing market for chemical fertilizers kept nitrate exports climbing steadily from 23,500 tons in 1850 to 117,315 tons in 1867.[14] The industry's expansion in turn prompted capital investment and technological improvement.

Refinement of nitrate technology took several different forms. The use of steam, rather than direct heat, to boil the caliche-and-water mixture reduced production costs and produced a higher grade of *salitre*. The introduction of steam-powered machinery for crushing the caliche further enhanced production efficiency. Nine *oficinas de maquina* (steam-powered refineries) incorporating the new technology were in operation by 1863.[15] But the structural weaknesses of the Peruvian economy slowed further development of the industry.

If Chile's economic history after independence was unique, Peru's constituted a classic story of the burdens of the colonial heritage. A central mining enclave of the Spanish imperial system, Peru emerged at independence with a declining mining industry and an agrarian sector dominated by traditional haciendas and corporate landholdings. Lacking adequate export products to fuel a system geared to the export trade, the Peruvian economy teetered toward collapse. An interest rate of 24 percent in Lima in the 1830s demonstrated the scarcity of liquid capital and the virtual nonexistence of financial institutions. Despite indirect ties to the capitalist economies of Europe during the colonial era, the wage labor mode of production had scarcely penetrated the country. The rural labor force was trapped on the great estates or clinging to village lands. In the cities, guildlike *gremios* maintained a tight rein over urban workers. Control of the country's labor resources by noneconomic means precluded a free wage labor market that capitalist enterprises could draw upon. The near total dominance of foreign merchants reflected a lack of domestic entrepreneurial skills to develop the local economy.[16] These structural deficiencies which inhibited domestic development of the guano trade, represented an even greater drag upon the nitrate industry where capital, technological, and labor requirements were much higher.

Until 1868 Peruvian mining laws allowed any individual to claim two *estacas* (one Peruvian *estaca* equals 20,795 square meters) of nitrate land. Individual members of a family or business enterprise would each register a claim in order to combine them into a single holding. This system permitted local interests, hampered by the scarcity of capital, to obtain huge land grants on which they established inefficient *oficinas de parada*. These undercapitalized operations created a virtual state of anarchy within the industry, as their owners abandoned them at the first adverse turn in nitrate prices, only to resume work once conditions improved.[17] An inadequate transportation system threw another roadblock in the industry's path of development. As of 1860, *salitreros* (nitrate producers) still used mules to carry their nitrate to the coast. Peru issued the first railway concession for Tarapacá in 1860, but construction did not begin until 1868.[18] Meanwhile, Chile's booming export economy rapidly created linkages with the Peruvian nitrate enclave.

After 1835 Valparaíso replaced the Peruvian city of Callao as the principal port on the west coast of South America. By 1842 Valparaíso had become the commercial center of the nitrate industry. Most of the ships returning from Tarapacá would anchor at the Chilean port to arrange for sale of their cargoes, and by the early 1870s three quarters of all nitrate sales were carried out in Valparaíso. Both foreign and domestic commercial houses became active traders in the product and extended credit facilities to the Tarapacá *salitreros*.[19] In 1872 an observer in Iquique placed the value of the nitrate trade to the Valparaíso merchants at 400,000 soles.[20] An important part of the trade was the provisioning of the Tarapacá *oficinas*.

Totally dependent on outside sources for food, Tarapacá emerged as a prime market for Chilean agricultural exports to Peru. In 1878 the Tarapacá *oficinas* purchased approximately 735,966 soles of Chilean agricultural products, or about 5 percent of Chile's annual agricultural exports.[21]

Chile also supplied a significant portion of the labor force as well as entrepreneurial and technical talent for the nitrate fields. Elements of the surplus rural population drawn to internal rail construction and mining enterprises were eventually attracted to *oficinas* and railway projects in Peru. Between 1868 and 1872 alone, some 25,000 Chilean peons migrated to Peru.[22] Chilean entrepreneurs such as Santiago de Zavala, Daniel Oliva, and Angel Custodio Gallo became involved in nitrate production in Tarapacá in the 1840s and 1850s.[23] Yet another Chilean, Pedro Gamboni Vera, adapted steam to the nitrate refining process and invented a method of extracting iodine from nitrate.[24] The growth of Chile's export economy thus enabled it to supply essential factors of production to Tarapacá while Peru still struggled with colonial structures that prevented development of its own resources. These conditions established effective avenues for further penetration by Chilean capital when the industry entered a period of intensified growth in 1869.

The decline of its chief competitor, guano, facilitated the takeoff of the nitrate industry. European customers complained of the inferior quality of guano that resulted from exhaustion of the richest deposits. Even high-

Table 1
Nitrate Exports from Peru and Average Prices in England, 1861–79

Year	Exports in Tons[1]	Average Price (£) per Ton[2]
1861	61,759	—
1862	74,046	—
1863	70,043	—
1864	49,572	15/5
1865	111,021	13/-
1866	99,440	10/19/2
1867	115,924	10/15/10
1868	86,659	12/13/4
1869	113,957	15/11/8
1870	133,790	15/10
1871	163,903	14/3/4
1872	200,943	15/-
1873	284,715	14/3/4
1874	253,783	12/5
1875	326,869	11/15
1876	320,491	11/11/3
1877	213,940	14/-
1878	268,601	14/10
1879	97,091	14/5

Sources:
[1] *South Pacific Times* (Callao), n.d. Copy encl. in Jerningham to Granville, Lima, 8 January 1873. F.O. 61/279; *PP*, 1878, vol. 72, "Peru. General Report by Mr. St. John," pp. 532–33.
[2] "Proposiciones relativas a la industria salitrera, informe del Señr. E. Malinauski a la Junta Consultiva," Lima, 16 September 1874. Copy encl. in GMS, 11,132; Miguel Cruchaga, *Guano y salitre* (Madrid, 1929), p. 173.

quality guano had only one half the nitrogen content of nitrate of soda, further enhancing the appeal of the latter in the European fertilizer market.[25] The lower production costs of the Gamboni system and the prospect of reduced transportation costs with the new railway further improved the industry's prospects. The immediate cause of the boom was an earthquake and accompanying tidal wave, which destroyed the shipping facilities at Iquique in 1868.[26] As a result, exports dropped by 25 percent that year, creating scarcity in Europe and driving up prices more than 25 percent in 1869 (see Table 1). The first great nitrate bonanza had begun.

Little more than a shabby collection of unpainted frame buildings scattered along unpaved streets, Iquique suddenly became a frontier boom town. A wave of humanity swept into the port, swelling its population from 3,000 in 1868 to 11,700 in 1871. Crowds stormed the offices of local notaries to register claims to nitrate properties. Men and women fought in the street over disputed titles to building sites in the town.[27] Chilean capitalists played a significant role in the now frenzied expansion of the industry.

The Caracoles silver discoveries in 1870, coupled with the prosperous condition of agriculture, provided a dramatic increase in Chile's domestic supply of capital. Credit operations of banks and other institutions increased from 12 million to 48 million pesos between 1866 and 1871.[28] In the midst of the economic upswing, a number of new joint-stock enterprises appeared, with mining companies leading the way. Nitrate corporations played an important part in this sudden burst of stock-exchange activity.

Thirteen nitrate companies were formed in Valparaíso between 1871 and 1873, twelve of them designed to exploit nitrates in Tarapacá. The one exception was the Compañía de Salitres y Ferrocarril de Antofagasta, a joint Anglo-Chilean venture, producing nitrate in the Bolivian department of Cobija that lay immediately to the south of Tarapacá.[29] Of the twelve remaining companies, two never advanced beyond the stage of paper corporations. The total authorized capital of the ten firms actually established, amounted to 3,700,000 pesos.[30] Their development was a direct result of capital scarcity in Peru and of Tarapacá's dependence on Chilean commerce. Creation of the firms took one of two courses, purchase of unused lands from owners with insufficient capital to work them, or buying out *salitreros* who were heavily in debt. The Compañía Salitrera Valparaíso exemplified the first type of operation.

In July 1872 six *salitreros* gathered in Iquique to combine their nitrate lands and form two companies, R. Olcay y Compañía and Cornejo y Compañía with total holdings of 1,000 *estacas*. One of the group, Oscar Herrera, assumed the task of obtaining financing for the companies in Valparaíso. Herrera was a logical choice, since as a merchant from the Chilean port he would have the connections needed to arange such an operation. By September, Herrera had accomplished his mission with the establishment of the Compañía Salitrera Valparaíso, which purchased 250 *estacas* from the Iquique firms. The company transferred half of its 500 shares to the six original *salitreros*, with the remainder held by new investors who provided capital to develop the property.[31]

The Compañía Salitrera San Carlos typified the second type of operation. In September 1872, a prominent Chilean, Francisco Subercaseaux, purchased the *oficina de maquina* San Carlos from Eujenio Marquezado for 340,000 pesos. Two thirds of the sale price went to pay off Marquezado's debt to the Compañía Chilena de Consignaciones y Depósitos, a firm in which Subercaseaux held a major interest. Subercaseaux then used the *oficina* to form the Compañía San Carlos in which his brother-in-law and business partner, Melchor Concha y Toro, served as a director.[32]

The leading figures in the new nitrate companies reflected the complex composition of Chile's upper social stratas. For example, Subercaseaux was the descendant of a noted landowning family. Major shareholders in other companies, such as Jorge Ross and Eduardo Délano, were scions of British and North American immigrants, involved in commerce and mining.[33] The importance of Chilean commercial ties to Tarapacá was evidenced in the predominance of the mercantile community in the makeup of the firms. Of the five companies for which data are available, 81 of a total of 111 shareholders identified themselves as *comerciantes* (merchants) or *dependientes de comercio* (merchant house employees).[34] However, neither establishment of the firms within a prosperous domestic economy nor participation of prominent members of the business community could guarantee their success.

While a dynamic export trade had opened the way for direct penetration of Tarapacá, the domestic economy contained formidable obstacles to the development of the nitrate enterprises. As Arnold Bauer has so admirably demonstrated, economic growth in nineteenth-century Chile "took place within a rural economy already staked out in large privately owned estates; where an abundant and underemployed mass of men and women could easily be set to work; and where a traditional landowning elite was the principal social and political force."[35] The response of agriculture to the export boom of the 1860s and 1870s clearly revealed the social productive relations of the Chilean *fundo*, or estate, which shaped the nation's social, economic, and political structures. Landowners achieved increased output largely through increments of land and labor. In the Central Valley, the heart of export agriculture, acreage devoted to cereal cultivation nearly quadrupled, while a population density of 14 per square kilometer provided an abundant labor force. Estates also increased the number of workdays required of their peons and anchored additional members of the floating population as service tenants, or *inquilinos*. Despite increased demands on labor, agricultural wages stagnated and fell behind rising food costs. Population pressures, the isolation of the countryside, hacendado paternalism, and the forces of religion served as effective non-wage-control mechanisms. Since labor had not emerged as a commodity that must be bought and sold in a competitive market, there was little impetus toward innovation. In the 1860s, estates in the Central Valley, valued at 100,000 pesos or more, used as little as 200 to 500 pesos worth of farm equipment.[36] Grounded in precapitalist (nonwage) class relations, Chilean agriculture responded to market incentives with additions of men and land as well as with increased exactions upon labor. In turn, it spawned a host of institutions that bolstered or compensated for its lack of innovation.

The national credit market, well developed by Latin American standards, geared its operations to the needs of the landed oligarchy, relying on land mortgages as the surest guarantee of loans. Financing of mining operations was generally restricted to commercial houses that handled their products. The higher risk represented by a foreign investment such as nitrates further limited the availability of such credit. The nation's rudimentary industrial base could not meet the needs of existing mining enterprises, compelling them to import most of their capital goods. An educational system that emphasized training in the humanities for the children of elite families did little to alleviate the shortage of technical skills.[37] As for marketing services, Chile depended on foreign merchant houses. Finally, the structure of the public education system reflected the domination of the state by landed interests. Although owners of large estates constituted less than 50 percent of the Chilean congress, the familial ties that bound the elite together gave them influence far beyond their numbers.[38] Thus, despite the existence of an effective centralized state, it was unlikely to serve as an agent in radically reducing such obstacles. Developments in the nitrate industry made it essential that the new companies overcome these problems.

After 1870, steady exhaustion of caliche deposits with the highest nitrate content and the rising cost of coal required constant upgrading and refinement of processing equipment. Technological improvements were also essential to deal with high labor costs. In addition to the limited availability of Peruvian workers, the migration of Chilean peons to Peru caused an outcry from landowners. Operating a labor-intensive system, Chilean hacendados feared that any reduction in the surplus labor pool would force up agricultural wages. Chilean peons were not formally enslaved or enserfed, and a steady rural–urban migration had begun by mid-century. Nevertheless, extension of the *inquilinaje* (service tenantry) system ensured retention of a rural population sufficient to man a labor-intensive agrarian system. Although the condition of the rural work force tended to depress wages throughout the economy and accelerate the process of migration, these factors were not sufficient to prevent labor shortages and the accompanying problem of higher wages that plagued nitrate producers.

With labor representing 50 percent of production costs *en cancha* (exclusive of shipping and handling) a solution was essential.[39] *Salitreros* responded with efforts to reduce real wages, such as by using *fichas* (company scrip). The labor scarcity itself limited such practices, since they were often met with mass resignations and work stoppages. *Salitreros* even considered the use of indentured Chinese workers to shield themselves from "the constant extortions of the Bolivian + Chili labourers."[40] But the most effective alternative was improved productivity.

Attempts to increase productivity included an acute division of labor that characterized *oficina* operations. Refining alone, exclusive of the extraction process, involved more that eleven different tasks, each assigned to a distinct group of workers. This specialization of function, in turn, necessitated the employment of qualified supervisors.[41] More significant in overcoming the problem was efficient technology.

The upgrading and refining of machinery to combat the declining quality of caliche and the high fuel and labor costs was based on adaptation of processes utilized in the European chemical industry. Therefore, importation of equipment and technicians from Europe became essential for successful competition in the industry.[42] Operation of the machinery required the employment of skilled labor such as machine tenders and chemical workers. Since these occupations respresented skill levels not readily available in Chile, they had to be recruited in Europe.[43] From their inception the Chilean companies attempted to reconcile conditions within the domestic economy with the requirements of the nitrate industy.

The capital of each firm represented the purchase price of the land and machinery or the funds required to build a new refinery. Operating capital had to be derived from one of two outside sources. The first, domestic institutions such as A. Edwards y Compañía and the Compañía de Consignaciones, were involved in merchant banking operations. These two institutions extended a total of 2.5 million pesos in credit to Tarapacá *salitreros* in the early 1870s.[44] Foreign merchant houses such as Schuchard and Company, and La Chambre Gautreau and Company served as the second source. In both instances credit was provided in the form of credit accounts bearing an interest charge of about 10 percent on which the nitrate companies could draw to meet their operating expenses. While these operations overcame the initial difficulty of capital availability, they did so at an interest considerably above the prevailing national rate of 8 percent. In almost every instance lenders tied these credit arrangements to consignment agreements. The merchant banking operation or commercial house handled the sale of the companies' nitrate, charging a commission of 1 to 3 percent for this service. Although the heady financial atmosphere of the period allowed the domestic institutions to compete with the foreign merchants in terms of interest rates, the merchant houses enjoyed direct ties to European markets. Such ties were crucial, since up-to-date, accurate information on prices and market conditions in Europe provided a major advantage in making nitrate sales in Valparaíso. Thus, although the impetus for the enterprises was Chilean, they were from the outset partially dependent on the foreign merchants who dominated Chile's international commerce.

Dependence on foreign sources extended to European products, technology, and technical expertise. The new nitrate firms had to import from England items ranging from nitrate bags and high-quality coal to complex refining machinery, engineers, managers, and skilled workers.[45] In this manner, the new companies strove to overcome the inadequacies of the national economy as they entered an increasingly competitive industry.

The ten Valparaíso companies were not the sole source of Chilean nitrate efforts. Several Chilean merchant houses and individual entrepreneurs also entered the industry as producers. The Chilean entrepreneur Daniel Oliva had been operating the *oficina* La China since 1860, and Jenaro Canelo worked an *oficina de paradas* (manually operated refinery). In addition, two commercial houses, Marcos Granadino y Hermanos and Soruco y Compañía, had purchased *oficinas* in Tarapacá. But, with 74 percent of Chilean productive capacity controlled by the joint-stock companies, Chilean success depended heavily on the fate of those firms.[46]

Initially, the companies displayed the signs of highly profitable enterprises. At the end of 1872, the companies reported their stock selling at premiums ranging from 5 to 60 percent.[47] The Compañía Pisagua recorded profits of 45,263 pesos after nineteen months of operation and paid dividends amounting to 22 percent in 1872.[48] By then, the firms faced intense competition from Peruvian and European interests.

Increased Peruvian participation in the nitrate industry resulted directly from the effects of the guano trade. The continued stagnation of Peru's domestic economy led members of the elite, including landowners and military men, to use their political power to syphon off revenues from the state-controlled guano trade. By the early 1860s, the interests of this group centered in financial and commercial enterprises in Lima.[49] A portion of their capital resources poured into the rapidly expanding nitrate operations of Tarapacá.

Between 1870 and 1874 six companies with a nominal capital of 3,620,000 soles were formed in Lima to produce nitrate in Tarapacá. At the end of 1872 the newly formed Banco Nacional del Perú established a branch in Iquique and, in the next two years, extended some 4 million soles in credits to nitrate producers. But the Peruvians as well as the Chileans were relative latecomers to the industry. Before the Valparaíso and Lima companies were even organized, European capitalists had significant investments in Tarapacá.

In the first three quarters of the nineteenth century, the Western European economies linked themselves to countries at the periphery of their system primarily through commodity exchanges, with the export of capital limited to financing the public debt of these nations. The increased export of capital for direct investment in transportation and raw materials production is generally attributed to the emergence of monopoly capitalism in the last quarter of the century.[50] In Latin America, however, European merchant houses had already initiated this transition.

The merchants of northern Europe had indirectly penetrated Latin America during the colonial period through their trade with the merchant guilds in Seville and Cadiz. When the Spanish empire began to crumble in 1810, these merchants, especially the British, rushed to fill what they perceived as an unlimited and now unencumbered consumer market. The acute concentration of domestic wealth that limited consumption of imports, and the initially unsuccessful search for export products to finance such trade, quickly put an end to these dreams. Forced to pursue alternative lines of endeavor, European merchant capital relied on marketing networks, capital resources, and eventually industrial capital's mode and means of production to achieve direct penetration of the Latin American economies.

Typical of such firms was William Gibbs and Company of Lima, a branch of the powerful London-based Antony Gibbs and Sons Ltd., which also had branches in Arequipa, Tacna, and Valparaíso. In 1842 Gibbs's marketing expertise and command of capital earned it a highly profitable guano consignment contract from the Peruvian government. Decline of the guano trade turned the house's interests to other investments, particularly nitrates.[51]

The event that brought Gibbs directly into the nitrate industry was a loan to a British *salitrero*, George Smith, in 1856. Nine years later Smith's indebtedness totaled 133,311 pesos. As a means of liquidating the debt, Gibbs entered into a partnership with Smith to operate his *oficinas*. The Compañía de Salitres de Tarapacá, formed in 1865, had three partners, Smith, Gibbs, and Melbourne Clark, an agent of Gibbs. By 1871 Gibbs became sole owner of the company.[52]

The Compañía de Salitres de Tarapacá, or Tarapacá Nitrate Company, soon developed into the most powerful nitrate concern in the Peruvian province. Its two *oficinas*, La Carolina and La Noria, accounted for 11.4 percent of nitrate production in Tarapacá in 1872. Although the Carolina was nearing the end of its usefulness, a new refinery, Limeña, was being built with an annual capacity of 800,000 quintals (one Spanish quintal equals 101.44 pounds). Its size made it the single most important *oficina* in Tarapacá before 1879.[53] The Limeña's capacity was twice as great as any Chilean *oficina* working or planned and possessed all the economies of scale involved in such a venture. In addition to the 622 *estacas* worked by its *oficinas,* the Tarapacá Nitrate Company (hereafter, T.N.C.) possessed 1,253 *estacas* for future development. Gibbs fashioned this commanding position within the industry out of its own capital and marketing resources.

Gibbs's investment in the T.N.C. totaled 1,316,172 soles in May 1873. The Lima and Valparaíso branches of William Gibbs and Company provided the company's chief source of financing. In turn, these branches could call upon the capital reserves of Antony Gibbs and Sons at an annual interest rate of 5 percent for operating capital and 6 percent for fixed investments.[54] Gibbs had also assembled a highly competent staff of British managers and engineers to oversee *oficina* operations. The T.N.C. marketed its product through the mercantile network built by the house in nearly fifty years of trading on the Pacific coast. Given the experience Gibbs had gained in the guano trade, it was a marketing system unmatched even by other European houses.

The other two major European nitrate companies originated in a fashion quite similar to the T.N.C. The British house of Hainsworth and Company with a branch in Valparaíso served as an *aviador* (creditor) to *salitreros* in Tarapacá. When one of its members, John Syers Jones, went to foreclose on overdue accounts in Tarapacá, he assumed ownership of the *oficina* San Antonio as a part of the debt collections. This property became the basis of the San Antonio Nitrate and Iodine Company, which Jones formed with two other British merchants, Campbell and Outram. Although the partnership underwent a number of name changes over the years, it was most commonly known as J. D. Campbell and Company. In 1873, Campbell's *oficina* had a productive capacity of 240,000 quintals.[55]

The third member of the European triumvirate, J. Gildemeister and Company of Bremen, Germany, served as a *habilitador* (credit and consignment agent) to *salitreros* through its Lima branch. When the 1868 earthquake left the firm with 500,000 pesos in uncollected debts, it foreclosed and entered nitrate production directly. The company's three *oficinas* possessed an annual productive capacity of 565,579 quintals in 1873.[56]

Two other significant European enterprises were the British firm Clark, Eck and Company, which began operations in 1873, and Folsch and Martin, a German partnership organized in 1872. They were distinguished from their counterparts by their dependence on outside sources for financing and consignment. Clark, Eck relied on Peruvian banks for financing; Folsch and Martin depended on the German merchant house of Vorwerk and Company for operating capital and consignment of its output.[57]

The same structural inadequacies of the Peruvian economy that facilitated Chilean penetration of nitrates had permitted European merchant houses to control a major portion of the industry. But similar means of access did not necessarily assure equal competitive strength to European and Chilean enterprises.

Table 2 provides a comparative analysis of the status of European and Chilean nitrate companies in mid-1873. In terms of the share of total industry production and productive capacity, the two groups appear to be close competitors. In terms of concentration of productive capacity, however, five European firms controlled nine *oficinas* with a productive capacity of 3,327,156 quintals versus a capacity of only 2,497,000 quintals for sixteen *oficinas* owned by fourteen Chilean companies. In other words, the average productive capacity of the European *oficinas* was twice that of the Chilean refineries. Furthermore, 85 percent of European capacity and 83 percent of European capital were in the hands of only three firms.

The superior productive capacity of the three largest European producers reflected an important advantage, internal credit resources at interest rates significantly lower than those available to the Chileans. They also enjoyed direct control over the marketing of their product. Up-to-date information on nitrate prices and market conditions in Europe, which such control assured, were vital for making profitable sales of nitrate.[58] The level of management and technical skills proved to be another critical area of differentiation.

The Chilean companies' indirect links to Europe resulted in an uneven pattern of personnel recruitment. Lax accounting procedures reflected the lack of experienced personnel. In fact, many of the companies did not know what their per unit production costs were. Personnel problems, in turn, led to frequent management changes.[59]

In contrast, a description of one of Gibbs's *oficinas* captured the essence of the efficient management characteristic of the European operations. "In the arrangement of the various operations there is a real vein of common sense running through all the departments; accurate estimates of the cost of every operation and every detail, are carefully recorded, and submitted for inspection each week to the manager at Iquique."[60] Furthermore, the

Chilean enterprises lacked the direct supervision over manufacture of *oficina* machinery provided for the European houses by their parent firms. The Antofagasta Nitrate and Railway Company offers further evidence of the important differences in efficiency which distinguished European and Chilean operations.

Unlike its Tarapacá counterparts, the Antofagasta Company used accurate cost accounting procedures and employed British technicians who overcame a series of production problems encountered by the firm. These distinctions stemmed from the fact that the firm was a partnership between the Chilean entrepreneur Agustín Edwards and Gibbs. Gibbs provided the company with a competent managerial and technical staff. In addition, the house's London office ensured quality control in the manufacture of the company's refining machinery.[61]

The advantages enjoyed by the European producers proved decisive, as developments in the early 1870s brought to light a fundamental contradiction that plagued the nitrate industry throughout the nineteenth century.

The growing market for the industry's product resulted from the accelerating demand for agricultural products in Europe. To meet the needs of the urban industrial complexes of Europe and the increasing per capita consumption of their residents, agricultural output intensified on the Continent, and new regions of the world such as the North American plains were put under the plow. Yet at the same time agricultural production methods in much of the world remained highly traditional. Furthermore, use of nitrates in mechanized agriculture proved impractical, closing off most of the immense market in North America. In contrast to the tradition-bound nature of its market, the tremendous input of capital and technology in the early 1870s converted the nitrate industry into an enterprise capable of rapid short-term increases in output. This contradiction of a traditional market fed by a modernized source of production intensified the frequency and degree of the crises of oversupply that characterized nineteenth-century economic enterprises.[62]

Between 1870 and 1873 exports grew from 133,790 to 284,715 tons (see Table 1). As supply outpaced demand, stocks of nitrate in Europe increased from 6,750 to 52,510 tons in the same period.[63] The average price of nitrate in England plummeted from £15 per ton in 1872 to £12.5 in 1874 (see Table 1). Complicating the problem for nitrate producers was a sharp increase in freight charges.

Initially, the German firm, F. Laisez, and the French concern, A.D. Bordes, dominated the shipping of nitrate. With the nitrate industry still in its infancy, the sailing ships of these companies were often unable to obtain a complete cargo of nitrate. In addition, the port facilities at Iquique

Table 2
Comparative Position of Chilean and European Nitrate Companies in Tarapacá, 1873

Factors	Industry Totals	Chilean Totals	Chilean Percentage of Industry	European Totals	European Percentage of Industry
Companies	115	14	12.17%	5	4.35%
Oficinas	143	16	11.90	9	6.29
Production 1872 (Quintals)	5,787,891.20	1,611,000	27.83	1,427,852	24.66
Productive Capacity 1873 (Quintals)	12,619,315.55	2,497,000	19.79	3,327,156	26.36
Average Annual Productive Capacity of Oficinas 1873 (Quintals)	88,246.92	156,062	n/a	369,761	n/a
Estacas	8,313	1,578	17.79	1,188	14.29
Value of Establishments (Soles)	6,117,418.89	1,349,500	22.59	1,768,240	28.90
Oficinas in Operation	38	9	23.68	8	21.05
Oficinas under Construction	26	5	19.23	1	3.89

Source: Table by Juan Ibarra, n.p., 31 May 1873. GMS, 11,129.

were still quite primitive. The sudden upsurge in nitrate exports strained the capacity of this fragile shipping network to its limits, and in 1873 freight charges jumped by 29 percent.[64] Just at the time when the industry faced a decline in the price of its product, the cost at which it could be delivered in Europe was on the increase.

Under these conditions, few *oficinas* could produce at a profit. In April 1873, James Hayne, the Gibbs manager in Valparaíso, reported that at the going price of nitrate, even the T.N.C., one of the most efficient producers in the province, could only break even.[65] Policy decisions of the Peruvian government further aggravated complications from falling prices.

The ephemeral prosperity of Peru's Guano Age was fueled not only by the state's guano revenues but also by extensive foreign borrowing. Even a consolidation of the internal debt was eventually financed through foreign loans. In 1869 the nation's external debt amounted to £8.6 million. To improve the darkening financial picture, the government of José Balta signed a contract with the French commercial house of Dreyfus Brothers and Company. In exchange for control of the guano trade, Dreyfus assumed the service of the country's foreign loans, agreed to pay off the state's 16 million sole (5 soles equaled £1) debt to previous guano consignees, and provided the state with a regular monthly income. Although bitterly opposed by Peruvian interests that had shared in the guano consignments since 1862, the Dreyfus contract did provide a degree of stability to government finances.[66] The state now initiated a massive public works program, primarily involving railroads, designed to ease the country's dependence on guano and to spur economic development. New foreign loans floated in 1870 and 1872 financed the program. The loans became the first milestones on a road that would lead the nation to bankruptcy in less than a decade.

By 1872 Peru's foreign debt stood at £35 million, and servicing the newest loans completely absorbed the state's monthly income from Dreyfus. Two factors made any improvement in the external debt situation extremely unlikely. First, the declining quality of guano and competition from nitrates had sent prices and exports spiraling downward after 1869. Second, the long period of extensive loans made to South American governments from the London capital market since 1848 was rapidly coming to an end.[67] The drastic undersubscription of the Peruvian loan of 1872 provided a clear sign of this trend. Prospects appeared equally bleak on the domestic side where 8.6 million soles (5.45 soles equaled £1) in nonguano revenues funded the state's internal expenditures of 17 million soles.[68] In September 1872, Peru's new president, Manuel Pardo, outlined a three-point program to deal with the crisis. The plan included fiscal decentralization, an increase in customs duties, and an export tax on nitrates.

Peru had imposed temporary taxes on the industry on five different occasions since 1828, but in this instance the Peruvian congress opted for an *estanco,* or state monopoly, of nitrate sales. The *estanco* offered the possibilities of increasing revenues, curbing nitrate's competition with the all-important guano trade, and appeasing small Peruvian *salitreros* who wanted the state to purchase their product at a fixed price. As promulgated in January 1873, the *estanco* called for production quotas assigned on the basis of output and capacity, with the state purchasing a maximum of 4.5 million quintals annually. The state would buy nitrate at 2.40 soles per quintal, and if nitrate sold for more than 3.10 soles, the state and the *salitreros* would divide the additional profits.[69]

Large producers, who may have perceived the recent price decline as an opportunity to eliminate their less efficient competitors, reacted sharply against the law.[70] They organized public protest meetings and refused to serve on a government commission to implement the *estanco;* a few even attempted to foment armed insurrection.[71] Their opposition led to an indefinite postponement of the law's enactment in September 1873. In place of the *estanco,* a fifteen-centavo per quintal export duty was imposed, but revival of the monopoly remained a distinct possibility.[72] This erratic course of events played havoc with nitrate prices. First word of the law's passage caused a sharp jump in prices, as buyers sought to avoid the higher prices ensured by the *estanco.* As its enactment became increasingly unlikely, prices fell to new lows.[73] Chilean nitrate producers were particularly hard hit by these developments, which were complicated by an adverse turn of events at home.

Chile's increasing dependence on world market conditions became apparent once again in 1873. A rapid drop in prices for silver and copper reverberated through the domestic economy as interest rates on bank loans increased from 10 to 12 percent.[74] Calls on the shares of joint-stock companies mounted to 6 million pesos, leading to depreciation of share values and further tightening of the money market. The recession, while not severe, eliminated the exceptionally favorable financial conditions that had allowed formation of the nitrate companies in an otherwise limited credit market. Furthermore, prospects of an *estanco* made investors particularly wary of the Chilean nitrate companies.[75]

Promulgation of the *estanco* law temporarily halted trading in nitrate shares as investors waited to see if the law would be enforced. When the entire market began to slip and nitrate prices continued their slide, shareholders scurried to unload their stock. Selling earlier at a premium, nitrate stocks by mid-1874 went for a fraction of their face value. Those who still held shares were unwilling or unable to pay quotas due on their stock. Rising interest rates and falling nitrate prices also caused institutional credit sources to dry up. The Compañía de Consignaciones and Edwards y Compañía halted further advances to the nitrate companies in 1874.[76]

Adverse financial conditions choked off attempts by the companies to expand their operations in order to cheapen production costs. An effort by the Compañía California to rent port facilities from the T.N.C., for example, was frustrated by high bank interest rates and the stockholders' refusal to supply additional capital.[77]

A further hindrance to the survival of the companies was the continued inadequacy of *oficina* management, which frustrated attempts at more efficient operation. Symptomatic of this problem were the negotiations between the Compañía America and La Chambre, Gautreau and Company. The French house agreed to supply America with operating capital but only on the condition that it be given control over the appointment of management personnel in Tarapacá.[78] Trapped between falling prices on the one hand and inadequate capital and management resources on the other, the companies began to disintegrate.

In January 1874 the Compañía Pisagua refused payment on 21,411 pesos in notes of credit written by its Iquique manager. Four months later, José María Necochea, its Valparaíso manager, was named to oversee the firm's bankruptcy proceedings.[79] Other Chilean enterprises were unable to pay their notes, collect from their debtors, or even pay their employees. To escape from mounting debts, the Compañías California and America tried unsuccessfully to sell out completely to their consignment agent and financial backer, La Chambre, Gautreau and Company.[80] The Compañía Solferino had utilized half of a 60,000-sole credit account from the Bank of London, Mexico, and South America to improve its *oficina*. Before supplying the remaining funds the British bank insisted on a new contract. The bank became sole consignment agent for Solferino. The company and the bank agreed to divide profits equally, with 5,000 pesos of the company's monthly share set aside to pay off its debt. Although new credit extensions were promised once the debt was settled, it was clear that the company was falling under control of the bank. By mid-1874 only four of the Chilean companies could keep their *oficinas* operating.[81]

The Antofagasta Company remained the one bright spot in the otherwise dismal picture of Chilean nitrate investment. The firm was not only free of Peruvian taxation and the threat of the *estanco,* but the machinery and technicians furnished by Gibbs enabled it to lower production costs by 20 percent between 1873 and 1876. In the same period, the company reported annual profits of 5 to 6 percent on its paid-in capital.[82] Meanwhile, the effects of the crisis were spreading and intensifying.

One observer in Tarapacá summed up the situation this way:

> Lately the impetus of new Comp's (in Peru + Chili) with their capital subscribed + habilitadores formed, had lasted, but now, in most cases at all counts, the capital is exhausted + the habilitadores will go no further—here many oficinas have either greatly reduced or altogether stopped production at losing prices.... most of the new works + Comp's are based on a capital barely sufficient for the purchase of grounds + establishment of oficina.... + have trusted for a floating working balance of about an equal amt, to other sources, which have for the present at all events dried up.[83]

The smaller European producers, also dependent on external credit, faced similar difficulties. Clark, Eck found promised loans being canceled and was bombarded with demands for early repayment of previous credits. As a result, the firm closed its refinery in August 1874. By October, Folsch and Martin had also halted production.[84] The larger European enterprises did not escape unscathed.

Negotiations for sale of the T.N.C. to Peruvian investors collapsed when the prospective buyers became leery of the proposed *estanco* and declining nitrate prices. Gibbs's desire to sell the firm early in 1873 represented an accurate reading of the company's immediate future. Between May 1873 and April 1874 the T.N.C. suffered a loss of 325,097 soles, most of it from the sale of nitrate.[85] Despite these setbacks, the European merchant houses weathered the crisis in far better fashion than their smaller rivals.

While the Chilean companies struggled to meet their operating expenses, Gibbs invested 500,000 soles to build the Limeña refinery.[86] In 1874 Gildemeister purchased the *oficina* San Juan and brought in German chemists and an experienced mining engineer to make it a model of efficient production.[87] At the same time, J.D. Campbell bought the *oficina* Agua Santa and increased its capacity.[88]

These firms with large, efficient *oficinas* already in production, and internal capital resources for expansion and covering temporary losses, could face the prospects of a continued decline in prices and even an *estanco* with a certain degree of confidence. Referring to the indefinite postponement of the *estanco,* the manager of the T.N.C., Henry Read, noted:

> looking at the matter from the point of view of a large producer, I think it is the best course that can be adopted, but to a small producer the delay will be ruinous.
>
> However it seems to me that whether the Estanco be established or not the same producers will have to succumb + it only becomes a question of whether they prefer being killed by inches or at once.[89]

As Read's words clearly indicate, the crisis in the nitrate industry differed significantly in its impact on two groups of producers.

By the end of 1874 the Chilean thrust into the Tarapacá nitrate industry had been blunted. Of the ten original nitrate corporations, two were bankrupt, five had stopped production, and the remainder were in serious financial trouble. In contrast their larger European rivals expanded and intensified their control of nitrate production.

Foreign penetration of the nitrate enclave up to 1874 was conditioned by the structural rigidities of the Peruvian economy. Peru's inability to make significant contributions of capital, labor, entrepreneurial skills, or marketing facilities permitted both Chilean and European capitalists to establish important linkages with the nitrate enclave by supplying these vital factors of production. This ability placed them in an ideal position to enter the industry directly as it began to boom in the late 1860s. Despite similar means of access, the Chileans were at a serious disadvantage from the outset.

Chile's position as a colonial backwater spared it the more severe economic dislocations of independence. The desperate search for export markets, the initial flow of cheap imported goods, domestic divisions over trade policies, and struggles to control land and rural labor were almost unknown in Chile. Its gradual absorption into the world market encouraged the development of a prosperous export economy and a cohesive elite with diverse but compatible economic interests. Political stability built on these factors was further cemented by the state's limited if growing role in both the domestic and the international economy. But incorporation into the world economy made Chile highly susceptible to the wild fluctuations common in the commodities market. Furthermore, growth of its export sectors had prompted only limited structural changes in the nation's traditional socioeconomic order. A social and economic system built upon a precapitalist, labor-intensive agrarian sector still hampered Chilean development. As a result, Chilean society was characterized by a limited capital goods capacity, an educational system as well as credit and labor markets geared to the needs of the landed oligarchy, and a state whose policies reinforced these institutions. The limits of this system became apparent in the Chilean effort to penetrate the nitrate industry.

The need for steady improvements in productivity in the nitrate industry imposed demands on the Chilean companies that could not be adequately met within the domestic economy. Chile's small industrial base and the scarcity of management and technical personnel required that such factors be imported from Europe. In turn, the inadequacy of *oficina* management reflected the firms' tenuous links to Europe. Nor did the companies enjoy direct control in the critical area of marketing. The Antofagasta Company, through its direct partnership with Gibbs, was the one firm that successfully bridged the gap between the limitations of the national economy and the constant need for increases in productivity in the nitrate industry.

In Tarapacá, the European merchant houses' superior command of capital and their direct access to the markets, technical skills, and industrial base of Europe allowed them to survive the crisis and accelerate their penetration of the industry. These developments were indicative of European merchant capital's new role as the conveyer of industrial capital's mode and means of production to the periphery.[90] Conversely, the failure of the Chilean companies exposed fundamental weaknesses in an economy that had achieved rapid growth in the context of a largely traditional social order. While direct Chilean interests in Tarapacá were collapsing, European control of the means of production increased. This line of development was unhindered and even facilitated by the Peruvian government's reentry into the industry in 1875.

[2]

The Peruvian Expropriation

THE decline of the guano trade after 1870 forced Peru out of an era of free-spending prosperity into a decade of unremitting crises. By 1875 President Manuel Pardo warned that social, economic, and political chaos threatened the nation.[1] To prop up the country's tottering institutions, the Peruvian legislature launched an expropriation of the foreign-dominated nitrate industry.[2] But the results of the nationalization were only partially successful. Although the surviving Chilean nitrate interests quickly shuttered, nationalization actually enhanced the position of the European merchant houses. From its inception, this program of nationalization was weakened by the fragile economic and political structures that had emerged in Peru during the Guano Age. As a result, the government relied on the power of the European commission houses to salvage its desperate attempt at expropriation.

The creation of a foreign-operated export enclave in the midst of a stagnant traditional economy had conditioned Peru's development since 1840. The domestic oligarchy controlled agriculture and mining but had been badly weakened by the decline of those sectors since the late eighteenth century. As a result, the state, which oversaw operation of the guano trade, emerged as the elite's mechanism for tapping a portion of the capital generated by foreign commercial houses. Beginning in the 1850s the oligarchy utilized its political power to enrich itself from state coffers. The consolidation of the national debt alone placed more than £3 million in their hands. State resources syphoned off in this manner became the basis for large commercial and financial institutions established by the elite in the 1860s. In turn, the principal endeavors of these institutions were the servicing of the guano trade, made possible by elite control of the state, and loans to the government.[3] Other methods of tapping state revenues emerged during this period.

Guano revenues and foreign loans secured with mortgages on the guano deposits financed the creation of an enormous military establishment and civil bureaucracy. A growing number of state pensioners added to the list of individuals dependent on the state for a major portion of their incomes. The government-backed railway construction projects of the early 1870s benefited the Lima banks that helped finance them and the commercial houses that provided materials for them. In addition, public works projects, especially the railways, provided employment for more than 20,000 workers.[4]

The extraction of state revenues by the elite thus created within Peru's traditional economy a "modern" sector of commercial and financial enterprises. Yet that sector depended totally on the guano resources of the state and on foreign loans secured by those resources. Since the government served as the primary avenue to wealth, political conflict became all the more intense, as factions of the oligarchy contended for the all-important revenues of the state.[5] European capitalism's penetration of Peru's traditional society thus created a system rent by two fundamental contradictions. First, there was a "modern" sector wholly dependent on a state that, by its very role in the process, was inherently unstable. Second, a supposedly autonomous state controlled an economy whose pace of growth or contraction was determined by external forces. By 1875 internal conditions necessitated continued economic expansion, while external conditions—that is, the decline of the guano trade and scarcity of foreign loans—dictated contraction. With the entire system now driven to the brink of collapse, Peru turned to its forgotten province for salvation.

Once the national congress focused its attention on the nitrate industry, it settled on a plan for purchase of the nitrate plants by the state. Although the plan appeared to be a drastic solution of questionable viability, there were valid reasons for its adoption. First, the fifteen-centavo duty on nitrates had produced only 912,551 soles in the period 1873-74. This would hardly satisfy the requirements of a state treasury that needed 30 million soles just to keep the railway projects functioning.[6] In order for the Peruvian state to continue to play its role of economic Atlas, it needed not only regular revenues but a new resource to mortgage against massive infusions of foreign credit. Second, a higher duty would harm the less efficient Peruvian *salitreros* who enjoyed a vocal representation in congress. Finally, the new project would lower nitrate production to improve conditions in the industry and limit its competition with the guano trade.

According to the scheme, the *oficinas* would be purchased for £4 million. Using the nitrate plants as collateral, the government would float a loan of £7 million in Europe. Of this amount £3 million would be employed in public works projects and the balance used to pay off the *salitreros*.[7] The project promised a return to prosperity, once again fueled by the credit facilities provided by a single state-owned resource.

The term "nationalization" in a twentieth-century context evokes images of grand schemes for national development. In this instance, however, nationalization was a desperate effort by an impecunious and unstable government to preserve the contradictory and now faltering system built upon a foreign export enclave. The entire nationalization process would bear the indelible imprint of this reality. Meanwhile, the expropriation, as it was termed, had become an inviting prospect for the *salitreros* of Tarapacá.

The crisis in nitrates continued unabated as the Peruvian congress debated the expropriation measure. After the first blows of the depression forced a reduction in output in 1874, those *salitreros* still able to find funds rushed to increase production in a hopeless effort to recoup their losses. In 1875 exports reached an all-time high of 326,869 tons, forcing prices down to £11.15 per ton, the lowest price since 1867 (see Table 1). The British minister in Lima reported in June 1875 that "competition is so great that lately nothing but losses have been incurred."[8] After examining the returns of the T.N.C. for May, June, and July 1875, even Antony Gibbs and Sons considered closing down its nitrate operations. Gibbs, however, still insisted it would accept only cash payments in any expropriation. Smaller and more desperate *salitreros* were willing to accept any type of note the Peruvian government might choose to issue.[9]

At the time of its formulation, then, the expropriation plan promised to rescue the complex of economic interests dependent on the state and ease the crisis within the nitrate industry. The British minister in Lima reflected this optimistic mood when he reported that:

> in the first place the manufacturers will be able to get rid of unprofitable works, next the Banks of Lima will receive back their heavy advances, and legitimate commerce will obtain that accommodation which it requires.[10]

With the promulgation of the expropriation law on May 28, these hopes seemed on the verge of fulfillment.

In September the position of the banks and the government was bolstered by a loan of 18 million soles (5.85 soles equaled £1) by the four principal banks of Lima to the government. The loan was guaranteed with various state revenues, including 4.5 million soles in anticipated receipts from the nitrate expropriation. The loan also aided local commercial interests by pumping currency into the economy as the state paid off some of its domestic creditors.[11] Meanwhile, the multiple interests dependent on the public works projects awaited the floating of the £7 million loan to guarantee the future of their operations. The fate of that loan reflected both the fragility of the state's attempt at expropriation and the adverse conditions in the London capital market.

Peru's extended performance on a fiscal tightrope had finally begun to unnerve its audience of European investors. The British minister in Lima specifically advised against further loans to Peru. In addition, after 1873 British exports to the Western Hemisphere, including loans to South American governments, underwent a sharp decline.[12] As it became apparent that the anticipated loan would not soon materialize, the paper currency issued by the banks began to depreciate. As a result, workers and pensioners found their incomes shrinking. With the state's fiscal crisis still unresolved, the salaries of the civil and military bureaucracy fell eight months into arrears, leading to renewed fears of violent outbreaks.[13] In addition, the public works interests now faced outright bankruptcy. Like two drowning persons, the state and the banks clung to each other in an ever tightening death embrace.

On April 29, 1876, the four principal Peruvian banks—Nacional del Perú, Perú, La Providencia, and Lima—signed a contract with the government by which they assumed direct responsibility for the nitrate expropriation. The four banks, now known as Los Bancos Asociados, would undertake the tasks of administering the expropriated *oficinas,* consignment and sale of the nitrate, and continuing attempts to float the £7 million loan.[14] The expropriation was a partnership of bankrupt institutions purchasing a severely depressed industry. The initial steps taken to launch the expropriation gave further evidence of the weakness of the administration that ran it.

The process initiated by the new law was an expropriation only in the most general sense of the term. Producers were guaranteed payment for their establishments in two years in the form of ninety-day notes on London. Meanwhile, they would receive 8 percent annual interest on certificates issued for the value of their properties. *Salitreros* who did not wish to sell their plants could continue producing, subject to an export duty that the government was empowered to raise to sixty centavos per quintal.[15]

In accordance with the law, a commission of engineers had been appointed in August 1875 to evaluate all the nitrate properties in Tarapacá. Once the evaluation was completed, a panel of lawyers would verify the titles of the properties before the final purchase agreements were signed.[16] The engineers accomplished their task in the astounding time of three months. It was later estimated that such an evaluation, if properly conducted, would have required five years to complete. Since the *salitreros* supplied the statistics, it seems certain that overevaluations were the rule and not the exception. Even with the survey completed, the Pardo government allowed some of the *salitreros* to delay the sale of their properties until they completed improvements, which further increased the sale price.[17] These actions prompted a contemporary critic to characterize the expropriation as the greatest giveaway in the history of a government renowned for generosity in dispersing its revenues.[18] They were in fact a reflection of the structural weaknesses of the Peruvian regime and the enormity of the crisis facing it. In the actual implementation of the scheme, the state depended heavily on the large European producers, who in essence held the future of Peru's government and "modern" economic sector in their hands.

Although many producers were still anxious to sell to the state despite the financial complications that had arisen, a number of points of resistance arose as the scheme went into operation. First, the Peruvian bureaucracy was hardly equal to the task of directly managing the industry. According to the plan, some *salitreros* would remain in possession of their plants after they had been purchased and operate them under state contracts. This was not a particularly inviting prospect for firms like Gibbs that wished to sell out completely and avoid the complication of production contracts. In contrast, the Associated Banks wanted to limit the contracts to firms such as Gibbs to ensure efficient production.[19] Second, Chilean *salitreros* felt particularly threatened by the expropriation scheme. With the memory of the *estanco* fresh in their minds, they viewed the expropriation as a new and more hostile attack on their interests by President Pardo. In the eyes of the Chileans, the new duty would be used to further reduce the value of their endangered enterprises so that the state could purchase them at a fraction of their real value.[20]

The Peruvian govenment, then, needed the cooperation of the large European producers whose command of capital and technology would assure efficient operation. These same firms could serve as stalking-horses for other *salitreros*.²¹ Furthermore, the state required the assistance of a large commercial enterprise capable of shipping and marketing its nitrate and advancing badly needed credits to the Associated Banks. Once again, one of the largest nitrate producers, the Gibbs house, most closely met the requirements of the feeble expropriation partnership.²² As if these advantages were not enough, Gibbs's partnership in the Antofagasta Nitrate and Railway Company in Bolivia raised the possibility that Peru could guarantee the success of the expropriation by limiting the company's output under pressure from one of its principal partners.²³ With the stage set for the crucial expropriation negotiations, the largest nitrate producers enjoyed a far stronger bargaining position than the Peruvian government.

Even before the law of May 28 was promulgated, Henry Gildemeister met with Henry Read, the manager of the T.N.C., to plan a joint strategy for negotiations with the Peruvian government. In view of the power of their firms, the two men concluded that they would demand production contracts totaling 2.5 million quintals per year. In addition, they would insist that the state allow them to produce the balance of the province' output at other *oficinas* purchased by the government. The two firms hoped to achieve a total monopoly of nitrate production in Peru.²⁴ The terms the two companies finally obtained did not constitute a total monopoly but were still highly advantageous.

In early May, President Pardo met separately with the senior partners of the T.N.C. and Gildemeister to discuss the expropriation. He explained to the Gibbs partners that he planned to offer them 990,000 soles for their property. They replied that their asking price was 1,250,000 soles; and "after much discussion and frank explanation he [Pardo] said he saw that it would be necessary to make an exceptional basis for us [Gibbs]."²⁵ Despite a later attempt by Pardo to reduce the price to 1,100,000 soles, both Gibbs and Gildemeister eventually received the higher figure for their properties.²⁶

Gibbs and probably Gildemeister received an inflated price for their establishments. At the time Gibbs did have 1,274,524 soles invested in the T.N.C., but the firm had suffered losses totaling 484,524 soles between 1873 and 1875. The continuing depression in the industry also made it highly unlikely that a private buyer could be found for the property.²⁷ The London partners concluded that "it cannot but be a matter of congratulations to feel there is good ground to hope you are about to be relieved of a business which was hardly likely to leave even interest on the capital employed."²⁸ In addition, Gibbs and Gildemeister were guaranteed 8 percent annual interest at a fixed exchange rate on the sale price of their properties and annual production contracts for 500,000 and 650,000 quintals, respectively, at a purchase price of 1.70 soles per quintal.[29] In light of these factors, Pardo's original proposal was far closer to the real market value of the properties. But the expropriation scheme, and thereby Peru's entire fragile economic system, depended upon the cooperation of these two large producers. Pardo had little choice but to acquiesce on the question of sale price.[30] The proposition was so attractive to Antony Gibbs and Sons that they declared themselves ready to go to any lengths, short of suffering losses, to ensure the success of the expropriation.[31]

With its interests now clearly defined, Gibbs moved quickly to shore up the shaky structure of the expropriation. It obtained the cooperation of other producers in carrying out the scheme and contracted for the consignment and sale of nitrate from state-owned *oficinas*. Although its primary interest in consignment was "to give a lift to the expropriation and so to the ultimate realization of the T.N.C. property," Gibbs nevertheless received highly favorable terms under the contract.[32] Its responsibilities under the agreement were limited, and it anticipated a profit of £3 for every ton of nitrate shipped. Far more important to the debt-ridden government and banks was the solution of the consignment problem and Gibbs's promise to advance the banks £50,000.[33] One form of assistance Gibbs could not provide was a limitation on the Antofagasta Company's output.

Initially Gibbs had been the dominant partner in the firm, providing the lion's share of the capital, as well as technology, marketing services, and management expertise. However, after increasing his fortune substantially by cornering the Chilean copper market in 1870, Agustín Edwards became the largest single shareholder in the company. Edwards also extended credits to the company that totaled 800,000 pesos by the end of 1876. With one *oficina* in operation, a second nearing completion, and duty-free export guaranteed by the Bolivian government, Edwards saw little reason to limit the firm's output. Efforts by the Gibbs partners to link a possible production limitation to an agreement for exclusive consignment of the company's nitrate to Gibbs did nothing to enhance the proposal's appeal in the eyes of Don Agustín. At one point Brice A. Miller, one of the Gibbs partners, wrote in exasperation that trying to bring the Peruvian government and Edwards together made him feel "like a matchmaker trying from interested motives to bring about a marriage between two unwilling people."[34] After the death of Don Agustín in 1877, the Edwards family continued his policy of noncooperation with the expropriation. When they finally agreed to limit the firm's output to 7,000 tons per year, it was only with the knowledge that it was the maximum the company's facilities could produce.³⁵ But this setback did nothing to dim the bright prospects Gibbs perceived in the expropriation.

Initially, the nationalization confirmed and even reinforced the dominant position of the two largest European producers. In August 1876, Henry Read confidently predicted that:

> the Nitrate business will fall into the hands of Gildemeisters, Folsch & Martin and ourselves until the Government increases the quantity to be exported.³⁶

The expropriation also confirmed the worst fears of the Chilean *salitreros*.

In accordance with the provisions of the law of May 28, the nitrate duty increased to thirty centavos per quintal at the end of that month and doubled again in December.³⁷ Still reeling from the nitrate depression, Chilean firms could either sell out and attempt to obtain production contracts or try to compete with the duty-free production of the contractors. During the nitrate crisis of 1873–74, Chile's traditional society had been unable to provide the corporations with the credit network, industrial base, or labor and management skills necessary for successful competition, causing them to accumulate enormous debts. As a result, the Chileans sought to secure production contracts. But the government had come to the conclusion that the nitrate duty would allow it to purchase nitrate from free producers at bargain prices, eliminating the need for additional contracts.³⁸ Furthermore, President Pardo had expressed the opinion that Chileans received an excessively large share of the profits from the nitrate industry. These factors frustrated the efforts of Francisco Subercaseaux and other Chilean *salitreros* to obtain favorable treatment from the president.³⁹ Any further hope of obtaining contracts appeared doomed by government decrees of July 6 and 13, 1876. The first increased the nitrate duty to 1.25 soles; and the second prohibited the banks from making any more contracts for one year. The latter decree also limited the nitrate purchases of the banks to 2 million quintals, including the 1,150,000 already contracted for with Gibbs and Gildemeister.⁴⁰ James Hayne noted at the time that:

> it appeared a matter of great doubt whether... any outsider [a *salitrero* without a contract] would venture to go on producing in the hope that the price would rise sufficiently to permit of his obtaining a profit at covering prices after paying so high a duty and in view of very heavy stocks on your side [England] and on the way, and that the Gov. Nitrate paid no duty.⁴¹

As the government's policies eliminated their viable options, the remaining Chilean joint-stock companies disintegrated.

In September 1875 the shareholders of the Compañía Valparaíso agreed to liquidate. As reasons for the decision, the directors cited the firm's debts, the general shortage of credit, and the export duties imposed by the Peruvian government.[42] Although the 200,000-sole offer by Peru was considered far below the market value of the property, the directors agreed to the sale to satisfy the firm's creditors.[43] Thus, the Valparaíso's demise stemmed from debts due to inefficient operation and the nitrate depression, inadequate credit facilities in Chile, and the effects of the expropriation. Suffering from the same problems, the Compañías America, Solferino, and California soon followed the Valparaíso into bankruptcy, with the certificates for their properties falling into the hands of their European creditors, the French mercantile house of La Chambre Gautreau and Company, and the Bank of London, Mexico, and South America.[44] The Pisagua, Sacramento, and San Carlos enterprises also wound up their affairs at this time, but the fate of their establishments proved to be distinctly different from that of the other Chilean firms. The threat of higher duties also forced a number of small Chilean *salitreros* to sell their plants.

By the end of 1877 both Daniel Oliva and Jenaro Canelo had sold their *oficinas* to the state. The expropriation had left small operators like Oliva with a return on their capital but no future in Tarapacá. As a result, they sought to set up nitrate operations in Taltal and Aguas Blancas, southern districts of the Atacama Desert controlled by Chile.[45] Prospects of the industry in these regions were limited. The deposits in the Atacama Desert were located farther from the coast than those in Tarapacá, and except for the properties of the Antofagasta Nitrate and Railway Company, they lacked the transportation and port facilities essential for profitable production.[46] While Chilean nitrate operators were being driven from Tarapacá, European investors were busy gaining control of the province' railway network.

In 1868, the state had given a concession to the Peruvian firm of Montero Hermanos to build a rail line between Iquique and the nitrate district of La Noria. The firm received a second concession in 1869 for a line from Pisagua into the nitrate region. These two railways were completed by 1875. Owing to the general scarcity of risk capital in Peru, the Monteros had secured two loans in London totaling £1,450,000 in order to complete the lines. They also transferred their railway concessions to a London-based company, the National Nitrate Railways Company of Peru, in which they retained a large interest.[47] By the beginning of 1875 the company had defaulted on the two loans. In accordance with the tems of the loan contracts, trustees for the bondholders of the first loan, and the Anglo-

Peruvian Bank, holder of the second loan, appointed an interventor to run the railway for their benefit. Although the Monteros retained a large share in the railway company, their continued indebtedness to foreign creditors made the railway a European enterprise, for all intents and purposes.[48] Meanwhile, Peru's continuing financial crisis contributed to another shift in government nitrate policy.

As his presidential term neared its end in July 1876, Manuel Pardo viewed the nitrate expropriation as a fait accompli. The purchase of *oficinas* had proceeded smoothly as most major producers had signed sales agreements. Both the decrees limiting nitrate purchases by the banks and the new export duty promised to keep state and private production at reasonable levels.[49] Lacking the foreign loan to complete the purchase of the *oficinas*, Pardo no longer intended to expropriate the entire industry, and a decree of July 13 halted the state purchase of *oficinas*.[50] Although exports remained high in 1876, they dropped dramatically the following year, and prices rose to their highest level since 1873 (see Table 1). Returns to the state from nitrates remained low, however, yielding only about £82,000 from August 1876 to July 1877.[51] Even if nitrate revenues had been twice this amount, they would not have been sufficient to stabilize Peru's shattered economy.

The national treasury was running a deficit of more than £1,250,000 for the period 1875 to 1876, and a deficit of more than £4.5 million was predicted for the fiscal biennium 1877–78.[52] In January 1876 the government defaulted on its foreign debt, and throughout 1876 and 1877 the banks continued to flirt with bankruptcy.[53] Henry Meiggs, the American entrepreneur responsible for building the railroads, was driven to issue his own paper currency to keep his operations afloat.[54] The only possible solution to the crisis was a foreign loan. Even before his election, the new president, Gen. Mariano Prado, had traveled to England in a vain attempt to secure the needed credits.[55] Once in office, Prado, like his predecessor, turned to expropriation as the only possible method of tapping the European capital market.

To obtain sufficient collateral for the loan, the president renewed attempts to complete the expropriation. A decree of November 29, 1877, authorized new purchases of *oficinas* by the state. By then, declines in Peruvian exchange and freight charges had lowered production costs. These developments, coupled with rising nitrate prices, had allowed companies such as J.D. Campbell to earn a profit as free producers. To bring these *salitreros* into the expropriation, the government had to offer not only purchase agreements but production contracts as well. Under this policy the Campbell Company agreed to terms.[56] The practice of utilizing the state to advance private economic interests also affected Peru's nitrate policy.

Clark, Eck and Company, only recently saved from collapse by a bank loan, received a contract. This fortuitous turn of events was the work of the father of one of the partners who was head of the Anglo-Peruvian Bank.[57] Otto Harnecher, a business partner of the president, received a contract for a state-owned *oficina,* as did Mescoso Melgar, an associate of Henry Meiggs and an agent for the Associated Banks.[58] The process of influence-mongering reached a level where even a few Chilean capitalists could preserve their interests in the industry.

Although the expropriation had spelled the end for the Chilean joint-stock companies, several wealthy Chileans attempted to rebuild their interests in nitrates from the ruins of these ventures. As the principal financial backer of the Pisagua and Sacramento enterprises, Agustín Edwards obtained possession of their properties when the firms liquidated. Edwards sold their *oficinas* to Peru and obtained production contracts for them.[59] Francisco Subercaseaux, the organizer of the San Carlos firm, purchased the company after it went into liquidation and then sold it to Peru in February 1878. Despite his failure at personal diplomacy with Pardo, Subercaseaux now worked to develop close relations with his successor and obtained a contract for his *oficina.* Aware of the inefficient management that had plagued Chilean nitrate companies, Subercaseaux placed the administration of the establishment in the hands of Folsch and Martin.[60] But the two contracts obtained by the Chilean entrepreneurs gave them control of only 7 percent of government production. By comparison, nine European firms controlled 52 percent of government output and thirty-five Peruvian firms held contracts for the remaining 41 percent.[61] Moreover, there was little chance that other Chileans would follow the lead of Edwards and Subercaseaux.

The Valparaíso business community had not forgotten the losses suffered as a result of the nitrate fever of the early 1870s. Commenting on Chile's economic situation in 1876, the noted economist Marcial González asserted that the only sector of the economy that had suffered appreciably since 1874 was that portion connected with Peruvian nitrates.[62] That same year a Valparaíso merchant argued that one of the most powerful causes of Chile's economic recession was the losses suffered in nitrate investment in Tarapacá.[63] Even Agustín Edwards concluded as early as 1874 that the real future of the industry lay in the Atacama Desert where it would be free from export duties.[64] Furthermore, alleged mistreatment of Chilean nationals living in Peru reflected tense relations between the two countries.[65]

But the bitter memory of earlier experiences and national animosities were not the most important factors deterring large-scale Chilean efforts in Tarapacá.

Chile's economic growth in the first three quarters of the nineteenth century had been conditioned by Europe's growing demand for foodstuffs and raw materials. Improved technology in shipping and the development of railroads allowed Europe to tap the resources of such far-flung regions as Chile. While providing a stimulus to Chile's agrarian and mining sectors, this same process opened alternative and, from a Chilean perspective, competitive souces of supply such as North America. These developments were reflected in declining prices for both mining and agricultural products on the London market. The price of copper, Chile's principal mineral export, fell from £91.10 per ton at the beginning of 1873 to £58.15 in January 1879. Between 1875 and 1880 the price of wheat dropped 13 percent. The labor-intensive structures of the Chilean economy were unable to respond to the price declines, thrusting the nation into the midst of its worst depression since independence.[66] National economic decline thus sealed off any further attempt to compete with European nitrate producers or to cope with the Peruvian expropriation. At this time, however, no position in the nitrate industry was an enviable one, since any semblance of a coherent government nitrate policy had disintegrated.

The distribution of nitrate production contracts to the influential was symptomatic of Peru's politicized economic system. Although placating a variety of domestic and foreign factions in the short term, it undermined the original intent of the expropriation—to create a profitable industry to be utilized as collateral for desperately needed foreign loans. During the Prado administration the state issued contracts for a total annual production of 8.5 million quintals. It was estimated at the time that this amount would have to be reduced to 3 million quintals for the industry to be able to operate at a profitable level.[67] This was reflected in James Hayne's observation that:

> The present position of Nitrate matters in Peru is extremely menacing to all concerned in Nitrate. The government in the most reckless manner has given out production contracts amounting to nearly or to between seven and seven and a half millions of quintals, and it may well be that we do not yet see the end of these contracts.[68]

Hayne no doubt feared not only overproduction but the disappearance of the monopoly once confidently predicted by Henry Read. There were good grounds for the latter concern.

With their original production contracts due to expire in 1878, both the T.N.C. and Gildemeister encountered serious difficulties in their efforts to renew them. The problems of the T.N.C. stemmed in part from Prado's attempt to use the contract as a lever in negotiating with Gibbs concerning their consignment agreement.[69] Moreover, Gibbs and Gildemeister were now simply two more contenders for contracts in a field that included other powerful foreign producers and a host of political and personal favorites of the president.

When Gibbs and Gildemeister did receive new contracts, they had to accept a 12 percent reduction in the purchase price of their nitrate. Both contracts were subject to cancellation by the government, and in November 1878 Gibbs's new contract was terminated. That same year negotiations between Gibbs and the government broke down, and the house lost the consignment contract for the state's nitrate.[70] Despite these setbacks, the expropriation had been highly beneficial to Gibbs and the other major European producers.

To counter the possible loss of their contracts both Gibbs and Gildemeister had purchased the contracts of other *salitreros*.[71] Even after the cancellation of the T.N.C.'s own contract the five major European producers alone controlled 36.03 percent of the contracts. By comparison, the 40.79 percent in Peruvian hands was dispersed among thirty-five firms, and Chileans held only about 7 percent.[72] These contracts were highly profitable for the producers.

Between 1876 and 1878, the T.N.C. reported total profits of 1,315,609 pesos, over half of which resulted from sales of nitrate to the government.[73] James Hayne concluded that:

> How important the expropriation and the Nitrate business generally is to WG + Co. is shown by their accounts for the 30th of April and 31st Decr/77. Without the gains from nitrate all arriving from expropriation measure—where should we have been![74]

Gibbs had also profited from its consignment contract and after its cancellation still made profitable nitrate sales in Europe.[75]

J.D. Campbell and Company, which had operated as a free producer until 1878, had not suffered serious harm. In the midst of the expropriation process the company invested 300,000 soles in its new *oficina*, Agua Santa. This *oficina* incorporated a new refining process that would revolutionize production methods in the industry.[76] Even the uncertainties of the expropriation worked to the advantage of European interests.

When the renewal of the expropriation failed to assure a foreign loan, *salitreros* who had not obtained contracts became concerned that they might never receive the full value of their certificates. As these producers began to sell their certificates on the open market, the uncertainties attending the expropriation sent the price of the certificates plummeting. For those with speculative capital available, it was a tempting opportunity. The Gibbs house purchased 306,000 soles in certificates at 55 percent of their face value.[77] The Bank of London, Mexico, and South America as the chief creditor of the Compañía Solferino received 600,000 soles in certificates issued for the firm's properties, and it purchased 160,000 soles of certificates issued for the Compañía America. Thus, these two British interests alone controlled 11 percent of the 20,297,901 soles in certificates issued by Peru as of 1878.[78] By the beginning of 1879 there was growing talk of returning the industry to private hands.[79] Possession of the certificates would be crucial in any such arrangement.

Under the expropriation the nitrate industry degenerated into a confused array of contractors without *oficinas*, *salitreros* without contracts, and certificate speculators with high hopes. In this uncertain situation only one group maintained a firm position in the industry, those *salitreros* who had received certificates and production contracts for their *oficinas*.[80] As Table 3 demonstrates, the important producers were limited to two groups, European and Peruvian *salitreros*, with the Europeans representing a far higher concentration of productive power and capital than their Peruvian rivals. Chilean interests, on the other hand, had been reduced to the point of insignificance.

Further European penetration of the nitrate industry under the expropriation was conditioned by the relationships that had emerged during the Guano Age and by the economic prowess of the European enterprises. The creation of a foreign-operated export enclave in the midst of a traditional society with a fractured elite and moribund economy established the state as the key link between capital generated in the enclave and the domestic oligarchy. Out of these relationships emerged modern commercial and financial enterprises dependent on state revenues and foreign loans. The fundamental contradictions in this system manifested themselves in 1875 as guano revenues and foreign loans evaporated. Under these conditions, the Peruvian government made a desperate attempt to reconstitute the system on the basis of the nitrate industry. The large European concern's command of capital, technology, and marketing facilities had already thrust them into a leading position within the industry. Their cooperation

Table 3
Oficinas Operating under Contracts and Held by Original Owners, December 1878

Companies	Sale Value (Soles)	Percentage of Total	Productive Capacity (Quintals)	Percentage of Total
5 European	3,990,000	50.38%	3,250,000	47.51%
35 Peruvian	3,529,191	44.57	3,150,000	46.06
2 Chilean	400,000	5.05	440,000	6.43
Totals:	7,919,191	100.00%	6,840,000	100.00%

Sources: *MH*, vol. 1091, Table no. 2; Read to Bohl, Iquique, 2 November 1878. GMS, 11,472/1; Cruchaga, *Guano y salitre*, pp. 289–92, 298–99.

was thus essential if the expropriation was to achieve efficient operation, thereby serving as adequate collateral for new foreign loans. To stave off collapse, the state was forced to grant generous concessions to this group, further strengthening their hold upon the industry. Even in this critical period, the use of the state for economic advancement undermined efforts to carry out a coherent economic policy as production contracts were dispersed among the politically influential. As the number of contracts multiplied, the hopes for efficient operation of the expropriation shrank. The politicization of economic activity made it impossible for the state to successfully construct even a dependent relationship with European capitalists, and the Peruvian system once more teetered toward collapse. Meanwhile, the twists and turns of Peru's nitrate policy and the structural inadequacies of the Chilean economy combined to deliver the coup de grâce to Chilean nitrate enterprises.

Their nation's inferior capital, technological, management, and marketing resources that had hampered Chilean *salitreros* in the early 1870s, and their lack of significant political influence in the Peruvian state denied them a strong bargaining position at the outset of the nationalization program. Subsequently, the deepening crisis in Chile's export economy precluded a resurgence of Chilean investment under the expropriation. Chilean ties to the nitrate enclave were increasingly reduced to the earlier pattern of commercial interchange.[81]

The failure of the Chilean nitrate enterprises by 1878 marked the limits of growth possible within the socioeconomic structures that had prevailed in Chile since independence. An export economy supplying Europe with raw materials through labor-intensive production methods and limited technological innovation had enjoyed nearly a half century of prosperity. The final collapse of the nitrate companies epitomized the decline of that system, as the development of competing sources of supply in the world economy threatened to propel Chile into an era of retrogression. Symptomatic of the limits of Chile's development in 1878, the nitrate industry would become the savior of that same socioeconomic order in 1879.

[3]

Elites and Foreign Investors: Chilean Nitrate Policy

BETWEEN 1879 and 1882, Chile went to war with Bolivia and Peru, gained sole possession of the nitrate regions, and instituted policies that governed the industry's operation for the remainder of the nineteenth century. Many historians have argued that the policies formulated at that time opened the way to the British nitrate monopoly of the 1880s. For a proper understanding of these pivotal decisions, it is essential to examine conditions in Chile prior to the War of the Pacific.

By the 1870s Chile's incorporation into the world economy had done remarkably little to alter the nation's fundamental social and economic institutions. The increasing number of merchants and miners who attained elite status did not effect a radical shift in the values of the oligarchy, nor did they prompt a thoroughgoing modernization of agriculture. Estate owners continued to respond to a growing export market with increased labor demands on their *inquilinos,* or service tenants, rather than with capital investment or production innovations. The lack of efficiency in agriculture and the failure of both mining and commerce to transform Chilean society resulted from the form of capitalist development in Europe as well as from indigenous conditions.

The process of industrialization in Western Europe in the first three quarters of the nineteenth century vastly expanded the region's productive capacity, stimulating increasing demands for raw materials. In tapping such sources of supply as Chile, the industrializing nations encountered domestically controlled production systems characterized by low productivity. The limited ability of such societies to improve their production methods contributed to a long-term rise in the price of raw materials down to 1873. Initially, the increasing cost of raw materials was more than offset by increases in labor productivity in Europe.[1] Thus, European capitalism's commercial interchange with these peripheral areas provided only feeble impulses toward restructuring their production methods. The history of the Chilean copper industry symptomized this phenomenon.

Despite a fivefold increase in copper production between 1844 and 1869, the industry continued to rely on production techniques from the colonial era until the introduction of the reverberatory furnace in the 1860s. Even then, the technological level of the industry remained low, its production methods labor intensive, and its transportation system inefficient. Where change did occur, it was achieved through inputs of British technology supplemented by the export of low-grade ores to European smelters. So too, the need for skilled workers was met by recruiting Cornish miners. Requirements for unskilled workers absorbed only a tiny percentage of the national labor force. As late as 1865 the northern mining region accounted for only 4 percent of the national population. This limited demand and the fact that the *inquilinaje* system placed no absolute check on migration of the surplus rural population precluded serious conflicts over the question of labor supply.[2] Thus, as intermarriage and the purchase of estates drew them into the oligarchy, mineowners had little reason to challenge such aspects of the traditional order as an inadequate industrial base or labor-intensive agriculture.

Domestic commercial interests supplied by European merchant houses with luxury items for elite consumption had even less motivation to promote structural change. Although they served as a link between a foreign system that emphasized productivity improvement and a domestic one that deemphasized innovation, this role in no way limited the profits they earned in handling exchange between the two. For example, Agustín Edwards as a merchant banker involved in the copper trade achieved his masterstroke not with innovation but simply by cornering the Chilean copper market and delaying sales until prices had risen sharply.[3] As long as Chile's exports commanded high prices on the world market, its merchant class was unlikely to challenge the existing order.

Chile's economic growth was taking place within traditional social productive relationships. These relations, relying on factors such as paternalism and religion rather than on wages as a means of control, generated strong disincentives to innovation. Labor was not entirely free—that is, a commodity to be bought and sold in a competitive market—so there was little or no compulsion toward productivity improvement, since output could readily be raised by increased impositions on labor. A premodern mode of production remained viable at a time when European capitalism's demand for raw materials was largely quantitative in nature and the production inadequacies of the societies supplying these goods were compensated for by increasing labor productivity in Europe. Within Chilean society this process facilitated the absorption of commercial and mining interests by the elite in the absence of fundamental conflicts over the structure of the productive process. In fact, Arnold Bauer has hypothesized that "much of the wealth made in the export enclave... simply ran out onto the sands of a traditional countryside."[4] This transfer of resources constituted an important regenerative mechanism for the landed oligarchy since low profit yields from agriculture and high expenditures to maintain the requisite life style "... returned little beyond ephemeral status and, frequently, downward mobility."[5] In addition, there is evidence of a crossover pattern. After 1870 as activity on the Valparaíso stock exchange intensified, landowners used their access to the national credit market to invest in commercial, financial, and mining companies.[6] In Chile, growth of the export sectors did not shift the balance of social or political power to a new commercial or mining bourgeoisie. Rather, the export sectors served to reinforce the structures of a traditional countryside that set the social and political standards of the nation.

Landowners facilitated this process by which dynamic export sectors bolstered the traditonal order through their control of a strong centralized state. The Caja de Crédito Hipotecario, established by the state in 1855, extended long-term, low-interest loans to estate owners. Although private sources of credit remained more important, the Caja's operations increased steadily, and the value of outstanding loans rose from 5,002,600 to 18,757,900 pesos between 1860 and 1880. Further assistance to agriculturists came in 1865 when the government offered to assume interest payments on all *censos* (liens) and *capellanías* (chaplaincies) pledged to support ecclesiastical institutions.[7] The northern mining rail network was left to private initiative, but the state developed an extensive rail system aimed almost entirely at tying the nation's agricultural regions to the port of Valparaíso.[8]

State support of agriculture was itself an indirect function of the export economy. Customs receipts, consisting almost entirely of import duties, normally provided 50 percent or more of the government's revenues. The state's fiscal position was a function of the nation's capacity to import, which was determined by the level of exports.[9] State assistance to agriculture and the continued absorption of mining and commercial interests into the landed oligarchy permitted Chilean society to maintain its essentially traditional character in the first three quarters of the nineteenth century.

Although expanding state services generally complemented rather than conflicted with the interests of landowners, certain points of resistance did develop. Rationalization of state functions led to fears of excessive disruption of the existing social order as government took an increasing role in domestic affairs. These concerns were expressed in such controversies as the church-state conflict over the secularization of education and eventually came to focus on the dominant branch of the government, the executive.[10]

Chilean presidents who were eligible for two consecutive five-year terms, and exercised a decisive legislative veto, successfully dominated the other branches of the central government. Efforts to limit presidential authority finally achieved some success in the 1870s when presidents were limited to a single five-year term and the electorate was expanded. Owing to presidential control of provincial and municipal governments, however, the executive continued to intervene successfully in the electoral process. This became apparent in the election of 1876 when President Federico Errázuriz imposed the innocuous Aníbal Pinto upon the nation as his successor. Pinto's administration (1876–81) was marked by further factionalization of elite political groups and intensified efforts by the legislature to curtail presidential authority.[11] But on the eve of the War of the Pacific (1879–83), the most immediate threat to the system was not political but economic.

After the recovery from the depression of the late 1850s, the annual rate of export growth underwent a serious decline from 7.2 percent between 1850 and 1860 to 0.6 percent between 1860 and 1870, and cyclical fluctuations in exports became a persistent problem.[12] Oscillations in exports were partially compensated by increased foreign borrowing. The state had secured only £3,246,300 in loans from the British between 1822 and 1866 but increased its indebtedness by £6,397,400 between 1866 and 1873.[13] Although with additional credits the government could pursue public works designed to benefit agriculture, the loans increased Chile's dependence on British sources of finance. The disintegration of Chile's position as a supplier of primary products was a far more serious development.

As raw materials came to represent a disproportionate share of the cost of production in European industry, the capitalist center strove to reverse the trend with improved production methods such as in the copper mines of the western United States and British copper mines in Spain. As a result, world prices for primary products declined after 1873. Falling prices for wheat, copper, and silver had increasingly severe repercussions within the Chilean economy. The continuing trade imbalance sent the metallic reserves of the banks flowing out of the country. Financial institutions found their debtors unable to repay their loans. In the case of agricultural mortgages, they were forced to accept repayment in land. The restriction of credit also hampered mining operations.[14] With more efficient producers entering the world market, Chile's export economy, grounded in a traditional society that deemphasized innovation, was reaching its developmental limits.

Although contemporary observers were alarmed by the nation's economic decline, recent scholarship has cast some doubt on the actual severity of what became known as the crisis of 1878. The decline of wheat prices was in fact far less severe than the drop in prices for copper and silver. Agricultural problems were in part transitory, resulting from bad weather and poor harvests. And it was not until the following decade that Chile's decline as a world supplier of wheat and flour became an undeniable reality. Furthermore, Chile's population was still predominantly rural and self-sufficient or, at best, marginally linked to the money economy. Reverses in the export sector would have a minimal impact on such a society.[15]

Such observations overlook the fundamental reality that while Chile was still demographically, socially, and politically an agrarian nation, it was mining that assured the persistence of this traditional order. As of 1870 mining, primarily copper, accounted for 60 percent of Chile's total exports; agriculture, only 30 percent.[16] Miners and merchants renewed the landed oligarchy with fresh infusions of blood and capital while landowners invested in the more dynamic export enterprises and the state tapped international commerce to aid agriculture. The decline of mining would mean the disintegration of the cornerstone in Chile's traditional socioeconomic edifice. By 1878, with copper prices down by 33 percent in only five years, that disintegration was shaking the fiscal foundations of the Chilean state.[17]

In 1877 the service of the foreign debt alone amounted to one third of budgeted government expenditures.[18] A series of attempts were made to offset the strain on government revenues resulting from the decline in export earnings. One project was an effort in 1876 and 1877 to encourage development of nitrate deposits in two Chilean-controlled regions of the Atacama Desert, Taltal and Aguas Blancas (see map, p. 6). Plans were made to develop ports and transportation facilities in these areas, and duty-free export was promised. These inducements, coupled with the Peruvian expropriation, encouraged a number of Chileans to initiate operations there, and by 1878 they had invested 4 million pesos.[19] It was clear, however, that such long-range developments would not solve the rapidly worsening economic situation. More orthodox solutions were tried. These included a 10 percent increase in import duties in 1877 and reductions in state expenditures. As government budgets for the period indicate, the general economic crisis and state indebtedness neutralized the effect of these programs. Although the budget was reduced from 4,025,673 to 2,750,367 pesos between 1875 and 1878, the state's revenues fell from 2,909,334 to 2,316,720 pesos in the same period.[20]

Writing to President Pinto in 1878, Alberto Blest Gana, Chile's minister in Paris, concluded that the crisis made it essential for the nation to reduce its dependence on mineral exports.[21] A few months later the minister consulted with the French economist Courcelle-Seneuil who had served as an advisor to the Chilean government during the 1850s. As a result of these discussions, Blest Gana recommended to Pinto a plan for taxing the incomes of Chile's upper classes. At the same time the minister noted that:

> It is true that with the policies that M. Courcelle suggests the banks and the monopolies will lose; it is true that capitalists will see their incomes threatened which until now have been more respected in Chile than animals among the *Hindus* who believe in reincarnation.[22]

As later events would prove, Blest Gana had accurately gauged the unwillingness of the oligarchy to alter a fiscal system that all but exempted them from direct taxation.

Finally, in desperation, the state secured a loan of 2,525,000 pesos from nine domestic banks. Since the banks were in no position to extend these credit facilities, the government granted them the privilege of having 10 million pesos of their notes accepted by the public treasury until 1888. The measure only aggravated the precarious condition of the banks, and by July the metallic reserves of one of the nation's leading banks were nearly exhausted. In emergency session the Chilean congress abandoned Chile's cherished tradition of hard currency and authorized the banks to suspend specie payments on their notes.[23]

It was generally agreed that the use of paper money could be no more than a temporary expedient, but, at the same time, congress rejected efforts to make any meaningful alteration in the state's fiscal structure. Only a week after the suspension of specie payments, six proposals aimed at taxing incomes and capital gains were submitted to the legislature. They were rejected out of hand.[24] Resorting again to traditional solutions, the government attempted to float a loan of £1 million in Europe. By December 1878,

Blest Gana reported that Chile's existing debt, the issuance of paper money, and unfavorable conditions in Europe made such a loan impossible.[25] Referring specifically to the issuance of paper money, one Chilean economist has accurately attributed these developments "to the weakness and contradictions 'in crescendo' of an economic and social system in decline, which sought to escape the old 'rules of the game' without being capable of a fecund and positive response."[26]

By the end of 1878, Chile's entire economic system was in the midst of its worst crisis since independence. Mining could no longer serve as the principal support of the national economy; orthodox remedies for economic problems had failed; and the oligarchy would countenance no dramatic alteration in the revenue-raising mechanisms of a state whose fate was closely tied to that of a precariously balanced banking system. At the same time, the increasing threat of war with Bolivia and Peru no doubt caused consternation among Chile's rulers. Yet the international conflict looming on the horizon brought with it a possible solution to Chile's declining fortunes.

The threat of war on the west coast of South America was in part the result of two international disputes that had troubled the region since the end of the independence struggle. Since that time Peru and Chile had competed for economic and political hegemony on the Pacific coast. When Chile's position was threatened by the Peru–Bolivia Confederation, Chile had gone to war in 1836 and destroyed the Confederation. By the early 1870s the rivalry between Peru and Chile had evolved into a race for naval supremacy.[27] This heightening of tensions was aggravated by Chile's economic penetration of Bolivia's Atacama Desert region.

Although historical precedent seems to support the contention that Chile's northern boundary was on the southern fringes of the Atacama Desert at about 25° south latitude, Chile had been claiming as far north as 23° south latitude since the 1840s. According to the terms of a treaty signed in 1866, the two nations agreed to share equally in the customs receipts from the region between 23° and 25° south latitude.[28] Left unsettled was the issue of sovereignty in the area and by 1872 efforts were under way to reach a new agreement. Meanwhile, Peru, no doubt alarmed by Chile's growing prosperity and increasing naval strength, urged Bolivia to reject any new accord, and the two nations signed a secret defensive alliance in 1873. Nevertheless, Chile and Bolivia agreed to a new treaty in 1874. According to its terms, Chile recognized a boundary at 24° south latitude and relinquished its rights to joint customs receipts. In return, Bolivia guaranteed that Chilean industries in the area would be exempt from new taxes for twenty-five years.[29] For the next four years international relations were characterized by an uneasy truce. Peru and Bolivia nervously watched Chile's penetration into the northern desert and its increasing naval strength. Chile, aware of their secret treaty, viewed her two neighbors with suspicion that was not diminished by Peru's expropriation of Chilean nitrate interests.[30] Then early in 1878 a new dispute over Chilean interests in the Atacama Desert set the three nations on a collision course.

The controversy was prompted by Bolivia's mounting concern over what it viewed as the denationalization of its desert province. Silver discoveries made in the interior of the region in the early 1870s had set off an influx of Chilean miners and capital. By 1874, 93 percent of the population of Antofagasta, the capital and principal port of the area, was Chilean. Bolivia faced the prospect of retaining control over a region that was becoming economically and demographically Chilean. Even its formal control was being challenged by Chilean secret societies whose goal was union with Chile.[31] In response to these developments the Bolivian National Assembly opted for a policy of confrontation by imposing an export duty on the Antofagasta Nitrate and Railway Company in February 1878.

The Antofagasta Company hoped to utilize its influential position within the Chilean state to counter the Bolivian threat. In addition to its president, Agustín Edwards, eleven of the company's shareholders were members of congress, including two who served in President Pinto's cabinet. However, several influential Chileans, among them the head of the Chamber of Deputies, Melchor Concha y Toro, and Senator Gerónimo Urmeneta, had Bolivian mining interests certain to be endangered in any confrontation with Chile's northern neighbor. To overcome this opposition, Antofagasta's directors launched a newspaper campaign presenting the company's cause as a matter of Chilean patriotism. The firm's efforts prompted Chilean diplomatic protests that delayed enactment of the export duty.[32] Finally, Bolivia's renewed attempt to collect the duty and its threat to embargo and sell the company's property led to the seizure of Antofagasta by Chilean naval forces in February 1879. In the weeks that followed, hurried negotiations occurred between Peru and Chile. Peru refused to abrogate its treaty with Bolivia and was finally drawn into the conflict. On April 5, Chile formally declared war on the two nations.

Clearly, a series of factors including geopolitical considerations and the influential position of the Antofagasta Company propelled Chile into the war. One of the most crucial factors was the rapid decay of Chilean mining, whose expansion had permitted the nation to experience significant growth and yet maintain its traditional social and economic order. This highly successful adaptation of a traditional society to its incorporation into the world capitalist system was now in jeopardy. It is by no means unusual for a society entangled in such a dilemma to seek a solution in expansionist policies.[33] The Chilean oligarchy was not blind to the possible solution to their problems offered by the nitrate riches of Peru and Bolivia. At the very least the economic decline had psychologically prepared the nation's governing elite to accept warfare as a possible escape from a crisis that other methods had failed to alleviate.[34] Once Chile achieved some early military successes, the dismemberment of Peru became an accepted policy to assure Chilean political hegemony. And within weeks of the capture of Tarapacá in November 1879, Chilean newspapers began to discuss the means of utilizing the new nitrate wealth to the nation's best advantage.[35] From that point on the war was most certainly fought for both political and economic objectives. From the moment the first Chilean soldier set foot in Tarapacá, however, the conflict's economic significance radiated out beyond the west coast of South America.

On the eve of the war, the continuing disintegration of Peruvian finances had caused serious international repercussions. European investors held some £30 million in bonds issued by Peru to cover its foreign loans. Of this amount, British subjects held between £24 million and £25 million. When Peru defaulted on its international debt in 1876, the bondholders organized a committee in London to protect their interests. With the outbreak of the war, the bondholders came to view Chile as a possible champion of their cause.[36] In August 1879, John Procter, a representative of the bondholders, was already in Santiago seeking Chile's help in protecting the bondholders' claims against Peru.[37] Specifically, Procter was interested in the Peruvian guano deposits mortgaged to the bondholders. By the end of the year, with, Chile in actual control of the guano and nitrate regions, the bondholders demanded permission to export guano. Of greater significance was the explicit claim made by the bondholders that Peru's nitrate works were also mortgaged to them.[38] And Chile had to contend with the Foreign Office as well. At the request of the bondholders' committee the foreign secretary, Lord Salisbury, instructed the British minister in Chile to ask the Chilean government to respect the bondholders' claims. In his instructions, Salisbury specifically repeated the claim that "the guano and nitrate of soda works now in the hands of Chileans in Peru have already been hypothecated to the bondholders."[39] Again on the request of the bondholders, Salisbury made mention of the nitrate works in a letter to Alberto Blest Gana.[40] By the end of January 1880, Chile had reached an agreement with the bondholders on sharing the guano revenues.[41] Since the British minister in Santiago had not presented his government's views on the bondholders' claims until January 29, Foreign Office pressure clearly played no role in the settlement.[42] But the claim to the nitrate works remained unsettled.

After learning of the guano agreement, the chairman of the bondholders' committee pointed out to Salisbury that Chile still had not made any mention of concessions on the nitrate of Tarapacá.[43] When a meeting of the bondholders accepted the guano agreement, it passed the following motion:

> That this Meeting is confident from the assurance of the Chilean Minister, that his Government will respect the rights of the Bondholders in regard to the Nitrates in Peruvian Territory.[44]

At this point, the Chilean authorities were apparently playing for time against the persistent bondholders. A few days after the bondholders' meeting in London, Miguel Amunátegui, the Chilean foreign minister, responded to the note of Francis Pakenham, British minister in Chile, concerning the bondholders' claim to the nitrate works. In answering the note Amunátegui simply restated Salisbury's views on the question without suggesting that he might be mistaken in regard to the mortgage on nitrates.[45] However tenuous the bondholders' claim might have been, there was good reason to avoid a confrontation at this time.

Chile was heavily dependent on Europe for war materials and future credit facilities. As the minister of finance noted:

> [the bondholders] have been a powerful lever in Europe preventing the Peruvians from acquiring war materials and creating for us a beneficent atmosphere in the opinion of those peoples.[46]

Chile was also acutely aware of what effect any dispute with the Foreign Office could have upon its war effort. After the seizure of seven Bolivians from a British ship, Pakenham had warned Amunátegui in January 1880 that such "recurring indignities to their flag" might cause the British government to enforce a strict neutrality on her subjects. As Pakenham noted, this would prove a serious problem for Chile, since most of its war supplies and even the transports for the army were in British hands.[47] Thus, Chile was barely in possession of Tarapacá when it was faced with an important foreign claim on the nitrate deposits. It was a claim Chile could not afford to ignore; nor was it the only one.

To finance its war effort, Peru attempted to arrange a series of international financial agreements. One of these was signed with the Société

Générale de Crédit Industriel et Commercial in London in January 1880. According to the agreement, the firm would take immediate possession of all government-owned nitrate works and share the nitrate revenues with the state and the bondholders.[48] Certain that the United States would be influential in any settlement of the war, the masterminds of the Crédit Industriel centered their activities in Washington. By January 1881 they had submitted a proposal for a United States protectorate in Tarapacá to the secretary of state, Charles Evarts. Their scheme was not as farfetched as it first might appear. In February 1880, the U.S. Congress had granted John Landreau, an American citizen, the good offices of the president and the secretary of state in pursuing his claim against Peru's guano revenues. This indicated a willingness on the part of the United States to involve itself in Peru's confused financial situation. Furthermore, the firm's counsel in Washington was the brother of the Speaker of the House of Representatives. Evarts showed some interest in the proposal and gave a letter of introduction to the company's agent, who was headed for Peru.[49] Thus, even before the new secretary of state, James Blaine, initiated a more aggressive United States policy on taking office in March 1881, "an economically motivated United States interest in the War of the Pacific had become apparent."[50] Chile's problems did not stop there. Those foreigners with the most direct interest in Tarapacá, the nitrate producers and certificate holders, made clear that they intended to have their interests fully recognized.

Saddled with a decaying economy and increasing war debts, the government was anxious to begin collecting nitrate revenues. While disruptions caused by the war made an immediate resumption of production impossible, there were 802,000 quintals of nitrate stored in Tarapacá ready for shipment. On December 26, 1879, an export duty of 1.50 pesos per quintal was imposed on all nitrate shipped from the province.[51] Led by Gildemeister and Gibbs, the *salitreros* refused to resume exports. They argued for a lower tax, claiming that the prevailing high prices in Europe would quickly drop once shipments began, making the tax an insufferable burden. Faced with this opposition, the minister of finance, Augusto Matte, dropped the idea of an export duty and ordered the producers to deliver nitrate under the terms of their original contracts with Peru. On December 6, 1879, however, the Peruvian government had issued a decree threatening contractors who delivered nitrate to the Chilean authorities with a fine equal to ten times the value of the nitrate. In light of this decree and fearing possible reprisals against their branches in Peru, the major foreign producers, including Gibbs, Gildemeister, and Campbell, refused to comply with the Chilean order.⁵² Tired of Gibbs's resistance, Chilean officials seized the firm's nitrate stocks on February 28, 1880; orders were also issued for the confiscation of Gildemeister's and Campbell's nitrate. Gibbs and Gildemeister lodged protests with their respective consuls.⁵³ Gibbs soon carried the matter beyond this initial stage.

On March 3, the Valparaíso branch of the house wrote to Francis Pakenham and protested the seizure. Pakenham in turn lodged a formal protest. A few days later, despite the protest, the govenment ordered the sale of the seized nitrate. After a visit from one of the Gibbs partners, the British minister presented another protest.⁵⁴ When his second note remained unanswered and the sale proceeded as scheduled, Pakenham sent yet another letter to the Chilean foreign minister stating that:

> I shall not therefore at present trouble Y E with any further observations upon what is, as I regard it, a very untoward affair; regrets are idle + remonstrance has proved to be in vain, but I will not attempt to conceal from Y E my belief that this enforced alienation of British property by the authorities of Chile will be viewed with profound surprise & concern by the Gov t of the Queen.⁵⁵

Miguel Amunátegui replied to Pakenham's note on March 23 and defended his government's right to seize the nitrate.⁵⁶

Gibbs now expanded its attack claiming, in a new letter to Pakenham, that if Chile wished to assume responsibility for the nitrate contracts it must also take responsibility for the nitrate certificates. And, as Gibbs went on to note:

> The term for the payment of these obligations having expired, they now constitute not only a Mortgage, but an overdue Mortgage with right of foreclosure accrued.⁵⁷

When in reply to Gibbs's latest protest Amunátegui made no mention of the certificates, Gibbs again wrote to Pakenham and insisted on recognition of the certificates.⁵⁸ For Chile to have recognized such obligations at that time would have been disastrous for the Chilean war effort. Already burdened with heavy war debts, the nation would have had to acknowledge a new debt of nearly £4 million.⁵⁹

This threat to its recently acquired wealth no doubt sent shock waves through the Chilean state. On May 3, Blest Gana appeared at the London office of Antony Gibbs and Sons with copies of Pakenham's correspondence. Promising a fair settlement of the dispute, Blest Gana asked the house to intervene at the Foreign Office on Chile's behalf. The London partners, however, informed him that it would be impossible to drop the protest. When questioned by Julian Paunceforte, permanent secretary of the Foreign Office, as to what course they wished the Foreign Office to pursue, the London partners advised him:

> that, for the moment, matters should not be pushed to extremes with Chili, as our Chief expressed it we desired that the sword of Damocles should remain suspended over their head but should not be allowed to fall.[60]

The strategy was a most effective one. When Gibbs pressed for a final settlement a year later, Chilean officials were so anxious to come to terms that they actually overpaid the house for the seized nitrate.[61] Nor was the state any more successful in its attempts to force Gildemeister and Campbell to resume production. Both firms managed to delay reopening their nitrate works until Chilean military victories had put their Peruvian interests out of danger.[62] Furthermore, other foreign interests had already taken up the question of the nitrate certificates.

By the end of 1880 the Bank of London, Mexico, and South America, held nearly 2 million soles in certificates on its own account or in the name of its customers and was seeking a settlement with the Chilean government. It wanted certificate holders to receive government bonds in equal exchange for the certificates.[63] Exclusive of the legality of such claims, the Chilean authorities had one very good reason for wanting to settle with the certificate holders. As one of the senior Gibbs partners pointed out:

> Probably a stronger weapon in the hands of the certificate holders lies in the fact that ere long Chili must be a borrower in Europe + unless the Certificate holders have been satisfied, it wd be very easy for them to throw insuperable obstacles in the way of her getting a penny— The Chilian Agents here know this + the readiness with which the Govt in Santiago consented to treat with Mr. Procter seems to show that they too are alive to the importance of dealing fairly by Peru's creditors.[64]

Chile then was in the midst of a terrible dilemma. Its armies had conquered a province whose resources promised relief for a beleaguered national exchequer. But that victory had also brought a deluge of claims by foreigners threatening that wealth and the country's primary source of foreign credit. Moreover, the incident with Gibbs had clearly indicated that the Chilean government could not simply dictate nitrate policy to the foreign producers in Tarapacá. In light of these problems, Chile opted to return the nitrate *oficinas* to private owners. This decision was preceded by the work of two commissions set up during 1880 and 1881 to study the nitrate question. In selecting the members of the commissions, the government predetermined the conclusion they would reach. Their members included the economist Marcial González and political figures such as Enrique Mac-Iver, well-known advocates of laissez-faire economic policies. Not surprisingly, both panels recommended the dismantling of the Peruvian monopoly. On June 11, 1881, the government issued its historic decree regarding the nitrate industry. By the terms of the decree any individual could obtain possession of an *oficina* by depositing with the Chilean treasury three quarters of the certificates issued for the establishment and paying the balance of its sale price in cash.[65]

Not all foreign certificate holders were satisfied with the decision. Many continued to press for even more favorable treatment. In a decree of September 9, 1881, the number of certificates required was reduced to 50 percent of the purchase price. Finally, a proclamation of March 28, 1882, gave the certificate holders definitive rather than provisional titles to the *oficinas* and ordered the auction of unclaimed *oficinas*.[66] Luis Aldunate, the minister of finance who signed this last edict, admitted some six years after the event that the government's motive in issuing it was a dual one. First, it wished to eliminate some of the nation's war debts; and second, in Aldunate's words, it wanted to escape from "an odious and tyrannical diplomatic claim."[67] By satisfying those with the most legitimate right to the nitrate of Tarapacá, the government could hope to silence at least one group of foreign investors and obviate the demands of those with less compelling claims. As the decree of June 11 stated, the *oficinas* were returned to certificate holders "without prejudice to the rights of third parties."[68] If other certificate holders, bondholders, or anyone else had claims on the *oficinas,* they would have to seek redress in the courts. The strategy proved to be an effective one. In 1882 a Foreign Office official concluded that the bondholders' claims on the nitrate deposits were impractical because the nitrate grounds were now in private hands.[69]

The foreign claims arising from Chile's conquest of Tarapacá exposed several weaknesses in the nation's economy. The expansion of exports along with the supply of factors of production and state financing from Europe had permitted it to experience impressive economic growth. But by 1878 declining prices for raw materials threatened to undermine this process. Furthermore, this course of development left the nation bereft of an adequate industrial base to supply its war machine when hostilities commenced in 1879. The economic crisis, preoccupation with preserving foreign sources of war materials, and protecting foreign lines of credit made Chile acutely vulnerable to European pressures on the nitrate question. These considerations were not the only forces that determined Chilean policy at this time. There were powerful groups within the country whose demands also shaped the fate of the industry.

With Chile still in the midst of its economic crisis, domestic economic interests had recognized the possible benefit to be derived from the nation's own nitrate deposits in Taltal and Aguas Blancas. As an article on the nitrate deposits of Taltal in the *Boletín de la Sociedad Nacional de Agricultura* pointed out:

> It is an incontestable fact that each advance each new improvement given to the industrial and economic life of the littoral, whether foreign or national, is translated immediately into greater prosperity for our agricultural production.[70]

The author also noted that the development of Taltal would open an important new source of commerce for Chilean merchants. As for northern mining interests, it was hoped that the railway planned for the nitrate industry would bring new life to the copper and silver mines of the area.[71] But by the beginning of 1880 the possible wealth to be derived from that region paled in comparison with the immense riches offered by the newly conquered provinces of Antofagasta and Tarapacá. Interest groups outside the executive branch of government sometimes outdistanced the executive in their desire to reap benefits from the windfall.

Soon after Antofagasta fell into Chilean hands, moves were under way to force the Antofagasta Company to pay for the assistance it received from Chile. Given the costs of the war and the state's precarious financial situation, such a measure was certain to enjoy popular support. But the prime movers behind the scheme had more specific economic interests in mind. Francisco Subercaseaux, who not only owned *oficinas* in Tarapacá but was a creditor and business partner of Folsch and Martin, was anxious to see the competition in Antofagasta saddled with an export duty. In the Senate, Subercaseaux could count on the support of his brother-in-law and business partner, Melchor Concha y Toro, and another in-law, Maximiano Errázuriz. Concha y Toro's position was particularly influential, since he served as head of the Senate finance commission that reported on the Antofagasta tax. In the Chamber of Deputies the campaign for a tax on the company was led by Ramón Barros Luco, an intimate friend of both Subercaseaux and Concha y Toro.[72]

The minister of finance, Augusto Matte, while not an investor in the Antofagasta enterprise, suggested a tax on its profits to avoid the more objectionable option of an export duty. But the company's political representation in the cabinet and congress was hampered by the nature of the threat. The issue at hand could not be couched in patriotic terms, as had been done with the Bolivian dispute. If Chilean politicians denounced and voted against a bill affecting an enterprise in which they had a direct interest, they would violate congressional regulations and leave themselves open to devastating political attacks.[73] With its political representatives limited to covert pressure, the company was able to defeat a one-peso per quintal duty proposed by Barros Luco, but it was unable to block passage of a still onerous duty of forty centavos per quintal.[74] This conflict was only a prelude to a far more important struggle triggered by the occupation of Tarapacá.

Soon after the capture of Tarapacá it became clear that, in the eyes of Chile's agricultural, commercial, and financial interests, the Peruvian province had replaced Chile's own nitrate regions as the answer to their problems. Agriculturists were quick to recognize the opportunity represented by Tarapacá. The initial attempt to impose an export duty on the province's nitrate was openly criticized because of its adverse effect on agricultural exports to the newly conquered area. The *salitreros* had refused to operate under the tax, and therefore:

> The Chilean agriculturist, who expected to find a market for his products finds the ports of Tarapacá closed because of an absolute lack of consumption.[75]

The effect of the capture of Tarapacá on Chile's sagging commercial sector was immediate. Reporting on the trade of Valparaíso for 1879 and 1880, the British consul there stated that:

> The foreign trade was at the same time conducted with caution, and as portions of the territory of Peru and Bolivia became thrown open to trade, and all the demand had to be supplied from Valparaíso, greater outlet was afforded for foreign imports as well as native produce, and commercial transactions, in spite of the repeated issue of paper money and consequent depreciation of that currency were satisfactory.[76]

Banking interests were also aware of the opportunities the Peruvian province offered. Within months of the capture of Tarapacá, the Banco de Valparaíso had established a branch in Iquique; and Chile's largest bank, the Banco Nacional, planned to begin operations in the city. The Banco Mobilario, owned by the Subercaseaux and Concha y Toro families, had been involved in extending credit to *salitreros* in Tarapacá even before the war.[77]

With Chile's export economy disintegrating, elite economic groups recognized that the newly conquered territories represented an alternative growth pole. Market linkages to the new nitrate region could provide an internal focus for the domestic economy without drastically reordering its production methods.

This new attitude toward Tarapacá became particularly apparent in 1880 when congress began considering an export duty for the entire nitrate industry. The resulting debates provided a clear exposition of the competing interest groups involved in the question as well as an indication of the attitude of Chile's ruling class toward the possibility of foreign domination of the industry.

The principal issue of debate concerning the nitrate duty was whether it should be applied uniformly to all areas under Chilean control. The idea of a uniform tax was vehemently opposed by the nitrate interests of Taltal and Aguas Blancas. They argued that since the newly opened regions lacked transport facilities and possessed inferior water supplies and nitrate deposits of extremely uneven quality, they could not hope to compete with Tarapacá under an equal export duty.[78] As a result of the various disadvantages of the southern regions, their production costs were some 40 percent higher than those of Tarapacá.[79] Despite the logic of their arguments and the fact that the interests of Chileans were at stake, the defenders of a proportional nitrate duty faced a difficult time in congress. The defense of the nitrate industry in Taltal and Aguas Blancas lay with a handful of congressmen from the northern mining areas whose constituencies included or bordered on the Chilean nitrate regions. So clear was the split between the north and the rest of the country that fears were expressed that a uniform duty would split the country socially and politically.[80] The only support for the northern representatives came from men like Justo Arteaga Alemparte, who espoused a protectionist economic policy on nitrate matters, and from representatives of the Antofagasta Company, which faced increased taxes if the uniform duty was enacted. In opposition were the majority of representatives from the central and southern regions of the nation.

The basic argument of the northern interests was simply that a uniform tax would destroy the southern nitrate areas of Taltal and Aguas Blancas. Enrique Mac-Iver, one of the principal advocates of a uniform tax, maintained that even if this should occur the general interests of the nation would not suffer, since the industry would be in the optimum condition of producing the maximum amount of nitrate at the lowest possible cost.[81] Mac-Iver dramatized the basic arguments employed by the proponents of a uniform tax when he stated that:

> The vote of the Honorable Chamber will decide if the already long and painful era of unbalanced budgets will come to a close, if the cancelled public works will be reestablished, if those factors which our situation and expedient

development demand will be created; or if the country will continue in this difficult and austere life which it has endured for four years.[82]

Expressed in blunter terms, the principal argument of uniform tax proponents was that the nation's desperate economic situation necessitated such a tax no matter what the consequences for the *oficinas* of Taltal and Aguas Blancas.

One of the most significant aspects of the debate was Justo Arteaga Alemparte's warning that a uniform duty would create a nitrate monopoly in Tarapacá, a monopoly controlled by "foreigners or enemies."[83] Even more to the point was the statement by Deputy Jovino Novoa, who declared:

> Nor should the Chamber forget, that, with few exceptions, the capital employed in this industry in Tarapacá, belongs to Peruvians or foreigners who do not reside in Chile, and that the capital employed in the establishments of the south belongs, also with few exceptions, to Chileans or foreigners residing in the Republic.
>
> Is it possible then to dictate a law that will mortally wound our capital and our industry?[84]

Novoa's statement was an accurate assessment of the problem, for at the time Chilean entrepreneurs dominated the industry in the southern regions but controlled only 6.43 percent of the nitrate production in Tarapacá, with the rest in the hands of Europeans and Peruvians.[85] But in reply, Mac-Iver simply stated that Chileans were represented in Tarapacá and that Chile could rely on the enterprising spirit of its people to prevent the industry from falling into foreign hands.[86]

These same arguments were repeated in the Senate. Carlos Walker Martínez reported that he had seen Peruvian documents listing owners of the Tarapacá *oficinas* and that not a single one was a Chilean. The owners were either Peruvian, English, German, or French.[87] The arguments in support of a uniform tax were presented by the Senate finance commission. The commission's conclusion in favor of an equal tax was, as the report itself stated, based on the ideas of one of its members, Melchor Concha y Toro. While Concha y Toro may have had the national interest as his primary concern, he certainly could not have forgotten that he and his brother-in-law, Francisco Subercaseaux, were committed to the success of Tarapacá through the Banco Mobilario and Subercaseaux's *oficinas*.[88]

Despite their accurate assessment of the conditions and prospects of the nitrate industry, the supporters of the southern nitrate interests constituted a minority in congress. As a result, a uniform duty of 1.60 pesos per 100 kilos was established by the law of October 1, 1880. Taltal and Aguas Blancas were later guaranteed 50 percent exemptions until June 30, 1882, and June 30, 1883, respectively, on the questionable assumption that these exemptions would provide time to build railroads in the regions and make them competitive with Tarapacá. Even if railroads had been quickly completed, the physical disadvantages of the southern areas and Tarapacá's long head start in developing the necessary economic infrastructure would have sealed the fate of the nitrate industry in those areas. Actually, they were unable to compete with Tarapacá until the late 1880s when an influx of foreign capital finally made them competitive. In the intervening years, the industry in Aguas Blancas totally collapsed, and only two of the twenty-one *oficinas* in Taltal survived.[89]

Chile's economic elite had been searching unsuccessfully for a solution to the state's unbalanced budgets that would not violate the traditional fiscal structure. A uniform duty on nitrates ideally suited their needs. It permitted the imposition of a high export duty without damaging the commercial, financial, or agricultural sectors. And the anticipated resumption of public works projects, especially involving railroads, would directly benefit their interests. The northern mining area would suffer to the extent that development of the region might be slowed by the decline of its nitrate sector. But even in the north only the Antofagasta Company and the new nitrate producers were directly affected. The north's principal economic resource was still copper and mining interests could take comfort from the prospect of solving the fiscal crisis without new duties on mineral exports. In fact, there was even hope that such duties would now be eliminated.[90] Furthermore, agricultural, financial, and commercial interests were willing to sacrifice the southern nitrate regions to benefit Tarapacá, since the Peruvian province promised the most immediate relief for the economic woes of these interests without restructuring the domestic economy. The desire to use Tarapacá as a stimulus to the economy was described in emotional language by one embittered opponent of the uniform duty when he wrote:

> To hope that the bases of the strong financial organization of the country would have been established in conformity with the principles of justice and equity, in accord with the mandate of the constitution of the State and affording equal respect to the interests and rights of all Chileans, would have been to delude oneself voluntarily and to attribute to the logic of lawyers, bankers and great landowners who make the laws in Chile and execute them at their whim, a degree of abnegation, prudence and patriotism which it has not manifested until now...

... nor was it natural that the circle of power brokers found themselves disposed to aid in the resuscitation of the national treasury and miss the opportunity of placing upon broad shoulders the full weight of the aggrandizement of Chile.[91]

In addition, Francisco Subercaseaux, one of the few Chileans with direct investments in nitrate production and financing in Tarapacá, obviously had a strong interest in seeing the competition of the southern nitrate zones crippled by a uniform duty. As in the case of the tax on the Antofagasta Company, Subercaseaux's interests had an influential spokesman in congress, Melchor Concha y Toro. Of course Subercaseaux's personal interests alone could not have been decisive, but they added a powerful ally to the ranks of those urging a uniform duty.

The majority of the economic elite, then, remained primarily concerned with the rapid redevelopment of Tarapacá to benefit both the state's fiscal position and to stimulate Chile's sagging economy. That this resuscitation of the Peruvian province would take place under foreign auspices precluded the necessity of radically restructuring Chilean society.

Chile's nitrate policy, which confirmed and ensured the powerful position of foreign interests in the industry, resulted from a convergence of foreign and domestic economic interests. The principal objective of the Chilean executive in its settlement of the nitrate question was to assure fiscal solvency threatened by a declining economy and increasing war debts. From the moment that Tarapacá came under Chilean control this objective was endangered by external threats. Peruvian bondholders, the Crédit Industriel, and nitrate certificate holders, with varying degrees of support from the governments of England and the United States, laid claim to Tarapacá's nitrate riches. The return of the industry to private ownership assured the Chilean state of a quick increase in revenues and removed it as the central target of the competing claimants. The policy also protected nitrate revenues from direct claims and prevented an immediate confrontation with foreign investors and their governments. The latter was particularly important if Chile was to avoid endangering its war effort and its principal source of foreign credit. The debate in congress concerning the nitrate duty made it clear that the policy had the support of the majority of Chile's elite.

Commercial, financial, and agricultural interests were all anxious to restore fiscal equilibrium. Equally important to them was the rapid resurgence of Tarapacá that would serve as a stimulus to the domestic economy. Only northern mining interests protested this affirmation of foreign control at the expense of Chilean nitrate producers. And despite this conflict, the entire elite could take comfort in the fact that a solution to Chile's worsening crisis had been found without abandoning traditional economic and fiscal structures.

Up to 1873 Chile had made a remarkably effective adaptation to its increasing interaction with the dynamic capitalist centers of Western Europe. Coming at a time of rising prices for primary goods, the nation's export sectors successfully responded with traditional methods of production, supplemented by limited infusions of European technology and skills. Growth was thus achieved without initiating fundamental challenges to the existing socioeconomic system.

This process of adaptation built on a commercial capitalist relationship with Europe had reached its developmental limits by 1878. By that time European capitalism was directly entering the process of raw materials production in far-flung areas of the globe. Its more efficient mode of production initiated a decline in world prices for primary goods that persisted through the end of the century. Chile's agricultural and mining industries, characterized by backward technology and labor-intensive practices, were unable to cope with the price declines. The innovation and productivity increases required to meet the crisis would have necessitated the restructuring of basic institutions, especially agrarian labor systems. The solution arrived at by the Chilean elite involved acceptance and promotion of a foreign-owned export sector within its national borders. The nitrate policies adopted at this time avoided the widespread societal changes required for domestic development of the industry. Such a decision might appear irrational by twentieth-century observers concerned with problems of development. Yet to an elite whose power flowed from the social productive relations of an archaic countryside the alternative course of innovation was an invitation to their demise as a ruling class. While preserving the traditional order, Chilean nitrate policies also shattered the clear division between domestic control of the means of production and European domination of international commerce and finance. This new relationship, once firmly established, would have dramatic consequences for the nitrate industry and the Chilean state.

[4]

Expansion and Foreign Penetration

AS the War of the Pacific dragged on into 1881, the Chilean government attempted, by means of a single proclamation, to extricate itself from a difficult international situation and return the nitrate industry to a position of stability and prosperity.[1] The nitrate decree of June 11, 1881, reinstituted private ownership by exchanging *oficinas* for Peruvian certificates, and it initiated what was perhaps the most sweeping and intense series of changes in the history of the industry. In little more than five years the Peruvian expropriation was completely dismantled. The industry's production technology was totally revamped, halving production costs and increasing annual exports to levels that were nearly double those of the previous decade. Begun under the reign of free market forces, this period ended with the first voluntary agreement to restrict production. Most notably, foreign control of the industry intensified, with European producers accounting for 75 percent of its output. These phenomena were the result of postwar conditions in the industry and the forces of competitive capitalism unleashed by the 1881 proclamation.

Successful execution of the nitrate decree proved to be a formidable task, since the industry was in a state of total disarray. Because of the haphazard enforcement of the Peruvian expropriation, the industry's system of ownership and operation was extremely complex. There were certificate holders working their own *oficinas* under government contracts, contractors without certificates administering state-owned establishments, owners waiting for their certificates, and certificate holders without contracts anxious to sell their apparently worthless bonds to waiting speculators.

The war and the Chilean occupation of Tarapacá had made the status of the province's principal industry all the more precarious and complex.

As Peru suffered a series of military reversals in 1879, it discontinued interest payments on the certificates. Initially, the Chilean seizure of Tarapacá aroused fears that both Peru and Chile might refuse to accept responsibility for the certificates. These developments further depressed the market for the bonds, and by the beginning of 1881, they were selling for as little as 11 percent of their face value. Not only did their price decline, but as speculators bought and sold nitrate bonds in Lima, the certificates issued for a single establishment were frequently dispersed among a number of individuals.[2] In Tarapacá the contending armies caused extensive damage to a number of *oficinas* and the nitrate railway.[3] Although Chile's armed forces had completely occupied the province by the end of 1879, doubts about the permanence of its control lingered until Chilean sovereignty in the region was recognized by Peru in the Treaty of Ancón in 1883.[4] In light of this confusion, and continuing United States interest in the settlement of Peru's financial affairs, Chile vigorously pursued its policy of private ownership in an effort to bring order to the industry and preclude foreign diplomatic intervention.[5]

After the provisions of the nitrate decrees had been complied with, many *oficinas* remained unclaimed, others were still being worked by contractors, and a number of owners were busy verifying titles to *oficinas* whose sales contracts with the Peruvian government had never been finalized. But with less than half of the province's *oficinas* in private hands by 1883, the private sector was nevertheless dominant. During that year, approximately 95 percent of the nitrate produced in Taparacá came from privately held *oficinas*. Moreover, many of the *oficinas* still controlled by the state were small and uneconomical.[6]

The return to private ownership brought with it the rapid increase in production sought by the government. Nitrate exports from Tarapacá rose to a new all-time high of 490,772 tons in 1883.[7] As a result, the export duty on the nitrate and iodine of Tarapacá produced 8,621,139 pesos that year.[8] These developments eased the international problems of the state and bolstered its financial position, but they also benefited two particular nitrate groups.

The most immediate beneficiaries of the confused conditions and efforts to rectify the situation were the European producers who had obtained certificates and contracts in the Peruvian expropriation. As Chile's attempt to enforce the contract system in 1880 demonstrated, these operators enjoyed a certain degree of security as neutrals and citizens of powerful European nations. In sharp contrast, their most immediate rivals, the

Peruvian *salitreros,* had fled before the Chilean advance. With the collapse of the Peruvian state and the financial institutions dependent upon it, few Peruvians attempted to resume operations.[9] For European producers, Chile's nitrate policy meant the immediate repossession of seven functioning *oficinas* with an annual productive capacity of 3,250,000 quintals.[10] The return of the certificates also allowed these companies to repossess the extensive tracts of unworked nitrate grounds they had acquired in the years prior to the war.[11] With these resources alone, Gibbs, Gildemeister, Campbell, Folsch and Martin, and Clark, Eck accounted for 31 percent of Tarapacá's production in 1883.[12] The extensive information on the nitrate region acquired by these firms through long years of experience was yet another competitive advantage.

The sense of uncertainty induced by the war, and the state-run auctions of *oficinas* and nitrate properties held in September 1882, made a number of establishments available at low prices. Many of the properties offered for sale either publicly or privately were old *oficinas* with worked-out grounds.[13] Identifying those establishments with actual or potential value required an expert knowledge of the industry. European producers possessed such information and had begun acquiring the most important properties soon after the Chilean occupation. Between 1881 and 1882 Folsch and Martin obtained more than 1500 *estacas* of nitrate land by private purchase. During the government auctions the firm also secured several small *oficinas de parada.*[14] Gibbs rented the *oficina* La Palma, with an annual capacity of 300,000 quintals, from the British-owned Peruvian Nitrate Company.[15] Since it was generally acknowledged that the long-range future of the industry lay in the untapped grounds of the province' southern districts, Gibbs also moved to obtain property in that area. During the public auctions the house purchased the Alianza, a tract of 340 *estacas* of rich nitrate grounds, located in the southern region.[16] These developments were accurately summarized by a group of certificate speculators who reported that those who benefited most from the uncertain status of the industry and Chile's nitrate policy were:

> the most powerful of the Old Nitrate owners acquainted with Nitrate manufacture, who aspired to free manufacture by acquiring the richest grounds and those best situated for production which were not yet in their hands, in order to subsequently destroy the second class producers by competition and so keep the monopoly of nitrate.[17]

These producers, however, were not the only ones who sought to take advantage of the confused conditions.

During the war, Lima became the site of intense activity by a diverse group of Europeans who speculated in nitrate certificates. Most of the speculators had little or no knowledge about the actual value of the properties represented by the certificates. Since they viewed the certificates as interest-producing bonds, which either Peru or Chile would redeem in hard cash, few of them bothered to secure a large number of certificates for any one *oficina*. The nitrate decreees left this group with the uninviting prospect of purchasing more certificates or combining their holdings with other speculators to launch ventures in which they had little or no experience. Therefore, most of them never attempted to exchange their certificates for *oficinas*.[18] There were, however, a few who possessed the expertise necessary to take advantage of the situation.

As the nitrate industry was transformed into a modern capitalist enterprise during the 1870s, the traditional societies of Peru and Chile had been unable to supply the management and technical talent required for its development. As a result, enterprising Englishmen such as Henry B. James and George Inglis had migrated to Tarapacá and enmeshed themselves in the economic life of Iquique. Others like James T. Humberstone, John Thomas North, and Robert Harvey found a ready market for their engineering skills. Prior to 1879 none of these men played a significant role as *salitreros*, but the instability brought on by the war allowed them to establish themselves as a force in the industry. The careers of North and Harvey, the two most successful members of this group, illustrate the process that brought these new *salitreros* to prominence.[19]

North first arrived in Tarapacá in 1871 after working for a British firm involved in the construction of the Carrizal railway in northern Chile. His first job in the province was at a Peruvian-owned *oficina*. The young engineer soon shifted the focus of his activities to Iquique. His most profitable undertaking was a company that supplied water to Iquique. During the war, North was left in sole possession of the enterprise when his partners fled the city, and he quickly established a monopoly in that vital service.[20] Prior to the war North became directly involved in the nitrate industry through his partnership in the firm of Brooking, James and Company, which purchased one of the production contracts granted to Agustín Edwards.[21] But it was the rapid depreciation of nitrate certificates that opened the door of opportunity for North and others like him. As neutrals he and other Europeans could invest in the certificates with a greater degree of security than the citizens of the belligerent nations. By 1879 North had developed a close friendship with Robert Harvey, and the two men became active on the Lima certificate market. Harvey alone was said to have made a profit of 200,000 pesos from these operations.[22] Obtaining more than simply speculative profits from the certificates required a precise knowledge of the value of the *oficinas* they represented. Harvey's experience in the industry made him and North something more than just two certificate speculators.[23]

Like North, Harvey had first worked in Chile's northern mining region and later emigrated to Tarapacá. In 1876 he secured a post with the newly created Inspección de Salitreras de Tarapacá. As a neutral, with detailed knowledge of the nitrate industry, Harvey was retained by the Chilean occupation forces as head of the agency.[24] The question that arose over Harvey's partnership with North was whether he contributed only his expertise or actually exercised his official powers in an unethical manner to advance their private interests. In particular, Harvey was accused of providing North with advance notice of the June 11 decree, thereby converting their certificate speculations into sound investments.[25] Although ethical behavior was not one of Harvey's distinguishing qualities, he was probably unable to provide North with more than a few months' notice of the decree, owing to uncertainty in government circles over what nitrate policy to pursue. Harvey's contribution, then, was probably a mix of inside information and his own expert knowledge of nitrate properties. Whatever information the two partners possessed, they employed it to purchase the certificates of a number of *oficinas,* including Jazpampa, Buen Retiro, Peruana, Ramírez, and Primitiva.[26]

North was also accused of using his friendship with John Dawson, the British manager of the Banco de Valparaíso, to gain unwarranted access to the bank's capital for his certificate purchases.[27] Although Dawson did serve as North's agent for some of his *oficina* purchases and later managed a bank organized by North, the evidence demonstrates that the charges are groundless.[28] The notarial records of Iquique indicate that almost all the loans to North were made after the June 11 decree and that in most instances North actually used his certificates or *oficinas* as collateral. In this manner, North obtained over 673,000 pesos in credits from the Banco de Valparaíso between April 1881 and November 1883. Furthermore, during the same period North obtained loans totaling 120,000 pesos from the British-owned International Mercantile Bank and another for 94,257 pesos from the ever cautious Gibbs house, indicating that he possessed sufficient collateral to obtain credit in a legitimate fashion.[29] With these credit facilities, North operated both the Jazpampa Nitrate Company and the Colorado Nitrate Company, founded in August 1881 to administer the *oficinas* Peruana and Buen Retiro.[30] North and Harvey were not alone in their efforts to establish themselves as new producers.

Henry Berkeley James and George M. Inglis took advantage of conditions in Tarapacá to create a foothold in the industry. During the 1870s James, as a senior partner in Brooking, James and Company, arranged credit facilities for *salitreros*. James saw the outbreak of the war as a direct threat to his principal business of nitrate consignment and credit. He felt that the Chilean invasion would be quickly followed by monopolization of the province' money market by Chilean banks.[31] Concerned for the future of his business, James expanded his direct involvement in nitrate production. James's junior partner, George Inglis, had experience in the management of *oficinas*.[32] James and Inglis combined their resources and formed partnerships with owners of nitrate grounds, providing capital to develop the *salitrero's* property in return for a share in the business.

Although three Chilean banks had now joined European merchant houses in providing operating capital to *salitreros*, neither group offered terms longer than twelve months. Thus, James and Inglis's offers of fixed capital investment proved extremely attractive to local producers, leading to the formation of J. T. Humberstone y Compañía in 1881, and J. Sanguinetti y Compañía in 1882. With their partners in Humberstone y Cía., the Italian *salitrero* Pedro Perfetti and the English engineer James T. Humberstone, James and Inglis developed Perfetti's nitrate property Tres Marías. Humberstone was included in the partnership, since his skills as an engineer would be invaluable in building the *oficina* planned for Perfetti's property.[33] The second company followed a similar pattern as James and Inglis allied themselves with another Italian, Juan Sanguinetti, to exploit his nitrate property, San José de Puntunchara.[34] Finally, in 1883 the two Englishmen formed a parent firm, James, Inglis and Company, to oversee their various nitrate interests.[35]

While much has been made of their personal virtues, or lack of them, these Englishmen's access to the industry stemmed from the inability of Peruvian or Chilean society to provide engineering and managerial talent for the modern industrial undertaking developing within their borders. Given the confused state of the industry, their experience in nitrates, and their relatively secure position as neutrals, these men were able to initiate nitrate ventures soon after Chile's policy had been announced. But for these new *salitreros*, and even the old manufacturers, simply obtaining and operating *oficinas* was no guarantee of success in the industry.

After 1880 *salitreros* still had to cope with many of the same problems that had confronted them in the previous decade. The need to resort to ever lower grades of caliche or nitrate ore was a continuing threat to productivity. In the short term, the demands of the Chilean war machine created scarcities of both fuel and labor. Even with the cessation of active hostilities, the cost of coal continued to rise.[36] The industry underwent a major technological transformation, with improvements in the refining process aimed at increased energy efficiency and treatment of low-grade caliche. But the high temperatures and sandy conditions of the nitrate region inhibited direct displacement of manpower with mechanical devices.[37] As a result, the industry's labor force increased from 2,848 in 1880 to a peak of 7,124 in 1882 (see Appendix). Despite the active recruitment of workers from the surplus population of central Chile, and the offer of increased wages, the nitrate producers encountered serious difficulties in meeting these labor needs even in periods of reduced output.[38] Although the cost of labor was low compared with such costs in the developed nations, the industry was still operating within a society without a completely free labor market. In the Chilean countryside, poor communications, strong family ties, religion, personal attachment to the land, and extension of *iniquilinaje* (service tenantry) limited the impact of a competitive wage system.[39] By 1886 the wages of nitrate workers were four times greater than those offered in agriculture. Higher food costs in Tarapacá and the use of company scrip lowered the real wages of nitrate workers, and payments in kind to peons also narrowed the gap (see Appendix). Yet the dramatic difference in nominal wages, which should have served as a highly effective attraction for workers, failed to have the desired effect. Thus, while the nitrate industry was operating in a society with a labor surplus, scarcity of workers and high wages remained serious problems for nitrate producers.

Under the Peruvian nationalization a number of inefficient *oficinas* had continued operations because of artificially high prices and because they had secured production contracts through political influence. In contrast, Chilean policies, by eliminating direct government control, freed the forces of nineteenth-century capitalist production to confront the problems of fuel, labor and the declining quality of caliche.

The Campbell Company had lured James T. Humberstone to work as a chemical engineer at its *oficina* San Antonio. While working for Campbell, Humberstone hit upon the idea of adapting a process used in the British alkali industry to the refining of caliche. The process, known as the Shanks system, gradually enriched the caliche-and-water solution by passing it through a series of boiling tanks before the solution was cooled and the nitrate allowed to crystallize. Since the new method permitted the exploitation of lower-grade caliche with greater energy efficiency, it lowered production costs dramatically.[40]

In 1878 the Campbell Company began building its *oficina* Agua Santa, which employed the new process. Four years later, the *oficina's* production cost *en cancha* (exclusive of handling and shipping charges) was sixty centavos per quintal, nearly 50 percent lower than the average of the *oficinas* operating in Tarapacá.[41] The system's enormous cost advantage made it essential that other producers adopt the system or drastically modify their own machinery to meet Campbell's competitive challenge.[42] *Salitreros* now entered a frantic race to build new *oficinas* (refineries), utilizing the system or introduce it into existing *oficinas*. A further step toward efficiency was a massive expansion in productive capacity. In 1878 the average annual capacity of *oficinas de maquina* (steam-powered refineries) in Tarapacá was approximately 241,000 quintals. By 1884 it had reached 666,000, and by 1886 it was nearing 777,000 quintals.[43] The widespread adoption of the Shanks system and the rapid expansion of capacity increased output per worker by 44 percent between 1882 and 1886 (see Appendix). Improved productivity reduced the average cost *en cancha* from 1.15 pesos to 68 centavos per quintal between 1882 and 1884.[44] The increased scale of production and improved productivity were the result of a long-term trend toward captial concentration. This basic characteristic of capitalist production, partly obscured by the Peruvian expropriation, had now burst forth with new vigor under Chile's laissez-faire policies.

In 1876, the Gibbs house had £152,908 invested in the Tarapacá nitrate industry. The capital of Gibbs's Iquique branch stood at £246,155 by 1882 and reached £385,696 four years later.[45] For that portion of its capital, borrowed from the London office, it continued to pay 5 percent interest on short-term loans and 6 percent on fixed capital investments.[46] These capital resources were crucial to Gibbs, since the grounds of its principal *oficina*, Limeña, were nearing exhaustion. The house answered the challenge of cheaper production, investing £64,000 in the purchase of machinery for a new *oficina*, La Patria. The new refinery, whose total construction cost reached £88,000, had a capacity of 1,156,000 quintals per year and a production cost of 42 centavos per quintal.[47] The case of Gibbs clearly illustrates the capital concentration necessary to compete in the nitrate industry. Its new plant had cost 36 percent more than the Limeña, the most powerful refinery in the province when it was built less than a decade earlier. Members of the Gibbs firm were well aware of the crucial role capital resources played in maintaining their position. In January 1885 one of the partners referred to:

> the prepondering influence which must fall to our lot as Owners of Oficinas capable of producing about Ql 3,600,000 pr. annum, with good grounds and above all commanding the requisite capital without leaning on the Banks or others.[48]

In a similar fashion, J. Gildemeister and Company had increased its capital from approximately £156,250 in 1876 to £249,166 in 1882, including £6,250 in improvements to its *oficina,* San Pedro.⁴⁹ It apparently devoted a good deal of its resources upgrading the machinery of its two largest *oficinas,* San Pedro and San Juan. Between 1882 and 1884 the cost per quintal at the San Pedro dropped from 1.30 pesos to 80 centavos, that of the San Juan fell from 1.10 pesos to 60 centavos.⁵⁰

J.D. Campbell had built its enormous refinery at a cost of £35,000 even while the industry and the Peruvian banks were suffering through the turmoil of the late 1870s. After the Chilean occupation, it received government permission to open the port of Caleta Buena as an export point for the nitrate of the Agua Santa. It then proceeded to invest £87,500 in developing the port and its transportation facilities.⁵¹

The fate of firms that did not command large capital resources for fixed investments is well illustrated by the case of Clark, Eck. The company had been in financial difficulty in 1878, and it was probably able to secure operating capital from the International Mercantile Bank only because the father of one of the partners was president of the bank. Despite its close ties to the bank, the firm still had to pay 9 percent annual interest on its credit. As the bank encountered financial problems of its own, Clark, Eck turned to the Banco de Valparaíso. By 1884 the company was forced to sell its *oficina* to settle a debt of 54,791 pesos to the Chilean bank.⁵² The high operating cost of its plant (1.60 pesos per quintal in 1882) created a major problem for the firm. But the root of its difficulties lay in the lack of sufficient internal capital to improve the refinery or to purchase and develop new properties.

Among the new producers, the partnership of James and Inglis was one of the few that could finance its development primarily from its own capital. With more than £60,000 of internal resources, James and Inglis built a refinery at Tres Marías with a capacity of 420,000 quintals and a cost of 60 centavos per quintal. Their other new plant at San José de Puntunchara had a capacity of 864,000 quintals and an operating cost of fifty centavos.⁵³ The situation of North and Harvey, on the other hand, typified the problems faced by most new *salitreros.*

North and Harvey had of course made extensive use of local bank credits. Some 470,000 pesos of these loans or credits were specifically earmarked for use in their nitrate ventures.⁵⁴ These short-term (three to six months) credits, bearing an interest of 9 or 10 percent, allowed the partners to operate their established *oficinas* of Buen Retiro and Peruana. But these credits did not permit the fixed capital investment required to develop their virgin grounds or refurbish their old refineries. Without such capital their careers in the nitrate industry were destined to be brief.

In 1881 Robert Harvey sought a solution to this problem. He wrote to Gibbs explaining that he and North lacked sufficient funds to build an *oficina* on their Ramírez property and offered the house the consignment agency if Gibbs would supply the necessary capital. When Gibbs rejected the proposal, North left for England in search of an alternative solution.[55]

During his career in Tarapacá, North had made the acquaintance of John Waite, a business partner of the Locketts, an important merchant family in Liverpool. The Locketts assisted North in answering the challenge of his competitors. On February 3, 1883, the Liverpool Nitrate Company Limited, with a paid-up capital of £100,000, was founded in London. Its leading shareholders included the Lockett, North, and Harvey families. Thanks to the Locketts, its list of stockholders was also graced by the names of other prominent Liverpool businessmen. Before the end of the year, machinery for the Ramírez *oficina*, the company's principal property, had arrived in Iquique.[56] By tapping the British capital market, North had secured the funds he so urgently required, and in less than five years this stratagem would radically alter the ownership structure of the industry. But a major reordering of the industry had already been achieved.

On the eve of the War of the Pacific labor-intensive *oficinas de parada* still played a large role in the industry, accounting for 26 percent of its output. With a quarter of the industry composed of these primitive operations, the average capital investment per *oficina* stood at approximately £23,000. By 1886 *oficinas de parada* had virtually disappeared as a production method, and average capital investment per *oficina* had reached £40,000.[57] The competitive forces of capitalism, unleashed by Chilean nitrate policies, had driven the industry to new heights of capital concentration to make possible the enormous increases in scales of production and productivity achieved in only five years. However, the process of capital concentration and the productive forces it created also threatened the industry with a catastrophic plunge in prices and profits.

By 1884 the race to expand and cheapen production had caused a 56 percent increase in exports and a 33 percent decline in prices in less than four years. In 1883 alone, Gibbs suffered a loss of £21,000 on its nitrate operations.[58] As production outstripped consumption, world stocks more than doubled, making a quick recovery from the depression unlikely (see Table 4). Furthermore, the greater concentration of capital within the enterprises made adverse operating conditions all the more unacceptable to their owners. However, capital concentration, while a root cause of the crisis, also created conditions favorable to its resolution.

The need to command large capital resources had resulted in a concentration of ownership. Just prior to the war, 123 companies had controlled 164 *oficinas* in Tarapacá. Six years later, 30 companies working 42 *oficinas* accounted for 97 percent of the nitrate produced in the province.[59] The limited number of operators, combined with the severity of the crisis, facilitated formation of the first voluntary agreement to restrict production. Beginning on August 1, 1884, the producers agreed to reduce total annual output by 20 percent to 10 million quintals. This figure was to be arrived at by quotas based on the capacity of individual *oficinas*.[60]

Since the Combination stabilized rather than increased prices, at reduced levels of output (see Table 4), it did not halt the forces of competition. This competitive struggle finally broke the Combination in 1886. By then it was clear that European producers had established effective control over the industry.

When the Combination collapsed, the six largest British and German *salitreros* accounted for 50.39 percent of Tarapacá's nitrate production (see Table 5). A diverse group of smaller European producers accounted for

Table 4
Chilean Nitrate Exports (in Tons), 1880–86

Year	Exports[1]	World Consumption[2]	Stocks[3]	Price in London[4] (£)
1880	223,974	n/a	131,000	15/10
1881	359,718	203,024.80	153,000	14/6/8
1882	492,246	290,335.20	256,000	13/5
1883	589,720	378,020.50	335,000	12
1884	558,900	496,202.40	351,000	10
1885	435,988	568,459,20	319,000	9/7/6
1886	451,030	441,881.80	323,000	10/5

Sources:
[1]Roberto Hernández Cornejo, *El salitre; resumen histórico desde su descubrimiento y explotación* (Valparaíso, 1930), p. 174.
[2]Chile, Ministerio de Hacienda, Sección Salitre, *Antecedentes sobre la industria salitrera* (Santiago, 1925), p. 41.
[3]Includes nitrate stored in consuming countries and in shipment from Chile. Chile, Ministerio de Hacienda, *Memoria de 1884* (Santiago, 1884), p. xciii; *Memoria de 1890* (Santiago, 1890), pp. lxxx–lxxxi.
[4]William Howard Russell, *A Visit to Chile and the Nitrate Fields of Tarapacá* (London, 1890), p. 373; *Statist* (London), 24 December 1889.

another 24.62 percent (see Table 5). Under Chile's laissez-faire regime, unrestrained competition had accelerated capital concentration to achieve innovation and expand production. This process increasingly placed control of the industry in the hands of the small number of producers who commanded the requisite funds for large fixed capital investment. But what had become of that enterprising Chilean spirit that Enrique Mac-Iver was certain would guarantee his nation's entrepreneurs a major share of the nitrate industry?

Table 5
Nitrate Production in Tarapacá, 1886

Producers	Number of Oficinas	Production in Metric Tons	Percentage of Total Production
6 Major European Producers	13	1,799,481	50.39%
13 Small European Producers	13	879,365	24.62
7 Others [a]	7	310,200	08.68
2 Chilean	6	264,819	07.41
5 Peruvian	6	241,407	6.76
Totals:	45	3,496,272[b]	97.86%

Source: *El Veintiuno de Mayo* (Iquique), February 1886 – January 1887.

[a] These seven *oficinas* were changing hands between small European and Peruvian producers after 1879, and it was impossible to determine in which group they should be included in the table.

[b] Tarapacá's actual total production in 1886 was 3,570,786 metric tons. The discrepancy in the table is due to the fact that a number of small *oficinas* produced for only one to two months during the year and therefore were not included in the table.

With European producers involved in a rapid expansion of their facilities, the direct Chilean investments that had survived the crises of the 1870s stagnated. Those limited interests were represented by three Chileans, Francisco Subercaseaux, Jenaro Canelo, and the Edwards family in the person of Eduardo Délano.

Eduardo Délano arrived in Iquique in February 1880 to take possession of the *oficina* Sacramento. Since the Peruvian government had never issued the certificates for the *oficina*, Délano repossessed it immediately and placed its management in the hands of George Inglis. Owing to the plant's high production costs, Délano eventually sold the *oficina* to Francisco Subercaseaux. Délano also purchased 548 *estacas* of nitrate land known as Lagunas from a Peruvian *salitrero*. Lagunas, like Gibbs's Alianza, was located in the undeveloped southern region of Tarapacá.[61]

While Délano seemed interested primarily in the industry's long-range future, Francisco Subercaseaux remained quite active in the industry's development immediately after the occupation. In the late 1870s, Subercaseaux had placed the operation of his *oficinas* in the hands of Folsch and Martin. After Chile occupied Tarapacá, this arrangement was continued and the Sacramento included in the agreement.[62] Far more important to Subercaseaux, however, was his control of the Banco Mobilario through which he continued to serve as a *habilitador* (credit and consignment agent) to Folsch and Martin and now extended credit to a number of other *salitreros*.[63]

For Jenaro Canelo the war brought a remedy for the paralyzation of his nitrate operations. Soon after the Chilean occupation he reopened his *oficina* San Fernando with financing from the International Mercantile Bank and the Banco Nacional de Chile. Like Clark, Eck, however, Canelo was working a relatively inefficient establishment and paying 10 percent interest for his operating capital. By the end of 1883, the International Mercantile Bank was pressing Canelo for repayment of his loans. Finally, to settle the debt he sold his *oficina* to Ceballos Sanz y Compañía, a firm organized by Gonzalo Bulnes, the former intendant of Tarapacá.[64]

Chilean *oficinas* accounted for only 9.96 percent of the total productive capacity in Tarapacá in 1884. Two of the Chilean *oficinas* were nearing exhaustion, and by the end of 1886 both had ceased operations. During that same year, Chilean plants refined only 7.41 percent of the nitrate produced in Tarapacá (see Table 5). The efforts of the three Chilean producers had been directed toward exploiting inefficient *oficinas* under the optimum condition of high prices that prevailed immediately after the occupation. Subsequent price declines further diminished the size of direct Chilean nitrate investment. It is noteworthy that even in these limited holding actions, both Délano and Subercaseaux relied upon foreign management, suggesting the continued scarcity of such talent in Chile. What their operations make apparent is that even Chileans with some experience in the industry made no serious effort to compete directly with the European *salitreros*. This dearth of direct Chilean investment can in part be attributed to conditions in the industry after the occupation.

European producers benefited both from their position as neutrals and Chilean policies concerning the nitrate certificates. But their ability to take advantage of these conditions was a direct result of the industry's development in the previous decade. During the 1870s the inability of either Peru or Chile to provide adequate capital investment or managerial and technical skills had prompted penetration of the industry by European merchant houses and an influx of Europeans to fill skilled positions. As a result, a small group of Europeans was uniquely positioned to develop the industry after 1880. Conversely, a lack of experience and expert knowledge served as deterrents to new Chilean investors. The free market conditions created by the state's nitrate policies placed even experienced Chilean *salitreros* at a disadvantage.

Chilean policies unleashed the forces of nineteenth-century capitalist production whose transformation of the industry had been distorted by the Peruvian nationalization. Declines in caliche quality, high fuel costs, and labor scarcity were met with intense capital concentration that enhanced productivity and concentrated the pattern of ownership. Requirements for heavy fixed capital investment, modern technology, as well as engineering and management skills, obstacles to Chilean investment a decade earlier, were now all the more intense. Such factors account in part for the paucity of Chilean investment, but they do not fully explain it.

While the forces of competitive capitalism were reaching their zenith in the industry, the potential for direct Chilean participation in nitrate production appeared to have been enhanced. The seizure of the nitrate regions ignited a period of spectacular growth within the domestic economy. Despite the improved conditions, new nitrate ventures did not appear, nor was there any widespread outcry for state policies to channel the new wealth into direct nitrate investments. The reasons for this apparent indifference derived from the impact of the industry upon Chile's traditional society.

[5]

The Symbiotic Relationship

CHILE'S seizure of the Peruvian and Bolivian nitrate regions in 1879 opened a new and unprecedented era of growth for the industry and the domestic economy. In the nitrate industry, European capitalists, freed from the restrictions of the Peruvian expropriation, quickly intensified the process of capital concentration. In the first seven years after the Chilean occupation they increased average capital investment per *oficina* by 74 percent, tripled average *oficina* productive capacity, revolutionized the means of production with the introduction of the Shanks refining system, and raised labor productivity by 26 percent. While European fixed capital investment and technology refurbished the nitrate industry, the export duty on nitrates fueled a rapid increase in government revenues, and private sectors of the economy benefited from links to the industry. Yet despite accelerated growth in the domestic economy, direct Chilean nitrate investments remained negligible. As early as December 1882 the distinguished statesman, Adolfo Ibañez warned the Chilean Senate that:

> According to data that I have at hand,... ten foreign houses alone have in their hands the necessary production to supply the world consumption of nitrate
> ...
> These ten houses can be converted tomorrow into five, afterwards into two and ultimately into one... Chilean [-owned] nitrate properties do not now exist in Tarapacá.[1]

But his warning went unheeded. Chilean entrepreneurs neither invested in nitrates nor expressed disapproval of growing foreign control.

As we have seen, wartime conditions and capital concentration magnified the difficulties for new entrants into the industry. Yet the absence of Chilean nitrate initiatives and the virtual sanctioning of foreign domination can best be understood in light of the impact of the industry upon Chile's socioeconomic structures. The capture of the nitrate regions reversed the economic decline of the late 1870s but required the nation to seek a new accommodation with the capitalist center. The direct implant of the capitalist mode of production within Chile's borders now overshadowed the commercial interchanges with Europe of previous decades. It is this relationship between the capitalist enterprises of the north and Chile's premodern society that underlay the tacit abdication of the nation's most valuable resource to foreign investors. The most important domestic element in this new relationship was the agrarian sector.

Prior to 1879 commercial agriculture in Chile was completely dependent on the export market. During the 1880s the decline of the previous decade accelerated. Annual wheat exports from central Chile that averaged 1,131,000 metric quintals between 1871 and 1875 fell to 86,000 metric quintals between 1886 and 1890. As the world price declined, the gross profit per *fanega* (1.58 bushels) of exported wheat slipped from approximately 1.10 pesos in the early 1870s to 10 centavos by the mid-1880s.[2] The incorporation of the nitrate zone, however, provided an alternative market. The development of Tarapacá illustrates the most direct tie between the nitrate industry and agriculture.

As an isolated desert province, Tarapacá depended totally on central Chile for subsistence materials. In 1884, 6,505 workers and 2,272 animals labored in the *oficinas* of Tarapacá. By 1890 these figures increased to 11,657 and 4,897, respectively. The province had imported approximately 1,182,000 pesos of Chilean agricultural products in 1878, accounting for 5 percent of Chile's agrarian exports. In 1887, 2,544,194 pesos' worth of such materials were introduced into Tarapacá. This figure equaled 27 percent of Chile's total agricultural exports for that year.[3] Occupation of the nitrate regions also signaled a more general growth in the northern market.

The population of Chile's Atacama province stood at 69,482 in 1875. With the incorporation of Antofagasta and Tarapacá, the total population of the northern region grew to 152,448 in 1885 and reached 201,475 by 1895.[4] Contemporary observers confirmed the significance of the northern provinces. In 1887 the British consul in Valparaíso reported that:

> Beans, maize, lentils, peas, dried fruit, & c., are seldom exported; the Chilean producer finding for these, as for his flour and barley, a better market in the northern desert region, where the mining and nitrate industries give employment to a large population, and require numerous animals, all of which have to be fed from the south. In the same way, the large and increasing wine and beer production of the south finds a market in the north for all its surplus.[5]

This same observer noted the compensating effect the region had for Chilean agriculturists whose wheat exports were declining:

> year by year the old irrigated lands between Santiago and Chillan are going out of wheat cultivation, and are being used for cattle feeding. This, the cultivation of vineyards on the slopes, and the production of barley and hay are found to give a better return in the northern and older districts than wheat-growing; the reason being the immense increase of consumption of feeding stuffs in the nitrate districts, where many thousands of mules have to be fed entirely on barley and hay imported from the south.[6]

The market for commercial agriculture was also on the increase in central Chile. Improvement in the national economy, prompted by the nitrate industry, accelerated urbanization, leading to a doubling of the domestic market between 1875 and 1895. Higher standards of living also increased per capita comsumption and the demand for a more diverse series of agricultural products.[7]

Expansion of the national credit network, where estates still served as the preferred form of collateral, further enhanced prosperity in the agrarian sector. Windfall profits for the nation's leading banks from short-term loans to nitrate refineries and related service enterprises, and state coffers bulging with nitrate duties, launched a dizzying upward spiral in the domestic money supply. In the new era of easy credit, one commercial bank added a mortgage section, and six new mortgage banks opened their doors. This optimization of credit facilities for agriculturists is best illustrated by the operations of the Caja de Crédito Hipotecario, still the single most important institutional source of agricultural loans. The total value of the Caja's bonds in circulation grew from 18,757,000 pesos in 1880 to 32,153,400 pesos in 1890. Equally important were the more favorable terms now offered by the Caja. In 1879, 79 percent of its bonds bore an 8 percent interest rate; by 1889, 71 percent of the bonds carried a 6 percent interest charge. At the same time, the Caja extended the repayment period of its bonds from twenty to twenty-four years.[8] Since most of the capital obtained from land mortgages was not reinvested in the countryside, these loans constituted a form of agricultural income. Summarizing the impact of the nitrate industry upon the agrarian sector, a Chilean economist concluded that:

> The rise of prices [for land], coupled with the decline of rates of interest, the expansion of credit, and the good market which the new nitrate provinces created for agricultural products, brought about a notable improvement in the status of the [agricultural] industry.[9]

Renewed prosperity in agriculture, however, did not signal sweeping structural changes. The increased availability of credit financed an ever more sumptuous life-style symbolized by the palatial homes that hacendados now constructed in Santiago. Estates expanded their use of land and labor, but as the National Agricultural Society noted in 1887, the proportional use of agricultural machinery was declining. Owing to differences in the rate of capital investment, labor use in agriculture was three times as high in Chile as it was in the United States for the same unit of output. The inability of agriculture to respond efficiently even to increased domestic demand quickly became apparent as food prices in Santiago increased 76 percent between 1879 and 1888.[10]

While capital investment in agriculture remained minimal, intensification of the *inquilinaje* system continued. *Inquilinos,* or service tenants, were required to work longer hours and supply additional estate workers, paying them out of their own pockets. New tenants added to the estates received smaller subsistence plots.[11] Although nonmoney payments make it difficult to trace trends in wage scales, all evidence points to stagnation in real agricultural wages (see Appendix). Ever expanding consumer and money markets triggered a new era of prosperity for a labor-intensive agrarian system, notable for its inefficiency and its contribution to the "grossly uneven distribution of income within the rural sector."[12]

The failure of Chilean agriculture to respond to new opportunities with capital investment is partly attributable to the limited capacity of market incentives to initiate such transformations. In this regard, the nitrate industry provided a disincentive to agrarian innovation, since it reduced the necessity of competing on the world market. But the primary obstacle to structural change was the structure of the hacienda and its role in Chilean society.

Within the confines of the hacienda, the landlord's control of his workers was assured by the forces of isolation, paternalism, religion, peasant attachment to the land, and fear of banishment into the legions of the surplus population that wandered through rural Chile in search of subsistence. Chilean agriculture constituted what Barrington Moore has described as a labor repressive system. The principal mechanisms of control in rural society were a host of "traditional relationships and attitudes." rather than competitive wage labor.[13] These premodern methods of control placed two effective roadblocks in the path of productivity improvement. Increased output could still be achieved by increased exactions upon labor, although this practice must eventually lead to declining productivity. More important, widespread innovation would undermine existing class relations.

Despite their diverse economic interests, the elite relied on the great estates to underpin their position. The estate was the solid building block of social status as well as of political and economic power in Chile. As we shall see in the next chapter, estate ownership served as an essential step in achieving social prominence for nouveau riche elements and in maintaining such status for established members of the oligarchy. More significantly, political influence flowed from the authority the hacendado exercised over the rural population. That political strength had recently been bolstered by the 1874 suffrage law that gave hacendados control of a majority of the electorate.[14] Although the precise mechanisms by which control of the rural populace translated into political power remain the subject of active investigation, its significance is apparent in the composition of the Chilean congress. In 1854, 41 percent of all congressmen owned rural estates. This figure increased to 50 percent in 1875 and reached 57 percent by 1902.[15] And while agriculture was characterized by low profitability, an estate was the key to unlocking the vault of the national money market. Rural property secured 80 percent of all mortgage loans made by financial institutions, and these loans went almost exclusively to large landowners.[16] Capital acquired in this manner could then be invested in higher-profit activities. The institutions of Chilean society were built upon an agrarian sector where a competitive wage system was of minimal importance in controlling the labor force. Preservation of this traditional rural society and its premodern class relations was thus essential to Chile's ruling class. In a society whose basic institutions had been shaped by the hacienda, decisions to restructure the social productive relationships of the countryside would have constituted a virtual abdication of power by the elite. The necessity of preserving the premodern class relations of the countryside so vital to the elite determined other responses to the nitrate industry. This is readily apparent in the policies of the Chilean state during the first decade of the Nitrate Age.

The state served as the most direct link between the industry and the domestic economy. Revenues from the export duty on nitrate and iodine soared upward from 6.9 million pesos in 1880 to 105 million by 1890, accounting for 52 percent of the state's ordinary revenues by the latter date.[17] These duties were the principal means of syphoning wealth from the nitrate region into the domestic economy. With ever increasing revenues at its command, the state, under the presidencies of Domingo Santa María (1881-86), and José Manuel Balmaceda (1886-91), expanded capital development investment. Education and public works became the focal points for these expenditures. The budget for the Ministry of Education nearly tripled between 1883 and 1890, and by the latter date, the budget of the Ministry of Public Works accounted for nearly 35 percent of all government outlays.[18]

Under Santa María, 1 million pesos were invested in new equipment for the state railways, and a 146-kilometer extension of the central rail line was approved.[19] Expenditures on transportation under Balmaceda included 1,200 kilometers of new rail lines and 1,000 kilometers of roadways.[20]

During Santa María's term, 700,000 pesos were spent improving port facilities in Valparaíso. Balmaceda's administration saw the construction of bridges, hospitals, and urban waterworks; the extension of telegraphic services; and the further improvement of port facilities.[21] These investments reflected the dramatic increase in the Chilean government's ability to develop the nation's economic infrastructure.

The Chilean educational system that before the war constituted "'a miserable' but a Chilean effort" enjoyed budgetary preference in the first decade of the Nitrate Age.[22] The total number of schools increased from 1,285 in 1881 to 1,653 in 1890. In the same period enrollment leaped from 79,930 to 150,000. Efforts were also made to upgrade and expand technical training.[23]

Increasing state expenditures were accompanied by expansion of the government buraucracy from 3,048 employees in 1880 to approximately 5,000 by 1891. Furthermore, salaries of public employees reached levels competitive with those in private enterprise.[24]

Contracts for public works, employment on those projects, improved transportation facilities, and increased salaries for the bureaucracy encouraged renewed growth in the domestic economy. Yet the question arises whether these efforts reduced the structural inadequacies that were so apparent in the failure of the Chilean nitrate enterprises of the 1870s. When examined more closely, there is little to suggest that they were either intended for or achieved such a purpose.

Although the government made significant outlays for infrastructure development, those expenditures must be evaluated in terms of their actual impact. Expansion of transportation facilities was largely aimed at improving the shipment of agricultural products to Santiago. The educational establishment increased in size, but heavy emphasis was still placed on the humanities. Furthermore the vast majority of the population was still excluded from effective participation in the educational process. In terms of supplying technicians and skilled workers, the system remained woefully inadequate.[25] These projects were the product of a state still dominated by an essentially traditional elite. Impressive in their dimensions, state capital investments served to reinforce that order which had been shaken by the crisis of the 1870s. Quantitative changes in the economic infrastructure that benefited the oligarchy's interests were entirely acceptable. But one could hardly expect members of this elite to launch initiatives that might threaten the existing order.

The manner in which the elite achieved economic growth while preserving and enhancing extant social and economic institutions is equally apparent in the response of the leading financial institutions to the nitrate industry. Large merchant houses such as Vorwerk and Company and

Williamson Balfour continued to be the preferred source of operating capital for *salitreros*. This preference was the result of their willingness to extend long-term credit (six to twelve months).[26] But Chilean banks filled an important role in providing short-term credit and arranging currency exchange operations. The financial institutions functioning in the nitrate zone included the nation's two largest banks, Nacional and Valparaíso, as well as the sixth largest, the Banco Mobilario.[27] They pursued a relatively simple method of operation in the nitrate zone. A *salitrero* would sign a sales agreement for a specified amount of nitrate with an export agent, or *corredor*. The time period allowed for delivery and payment of the nitrate was usually two months. The producer would then use the sales agreement to obtain sufficient credit (minus interest charges) from the bank to extract and refine the caliche. Once the producer delivered the nitrate the bank recovered its loan. These discounting operations allowed the bank to make short-term, high-interest loans. The banks frequently opened six-month accounts for their best customers to provide them with a regularized source of operating credit. Such accounts were guaranteed by a mortgage on the *salitrero's oficina* and, like the short-term bills, carried an annual interest rate of 10 percent. Some of the smaller *salitreros* paid interest rates as high as 2.5 percent per month for their operating capital.[28]

Chilean financial institutions also found other willing customers in Tarapacá. Firms dealing in the purchase and sale of nitrate kept the funds needed for their operations on deposit in the branches of the Banco de Valparaíso or the Banco Nacional. In 1883, for example, the Gibbs account in the Banco Nacional in Valparaíso stood at 485,000 pesos with another 206,000 pesos deposited in the Iquique branch of the Banco de Valparaíso.[29] The Banco Nacional, as the official depository of the Chilean exchequer, received all payments for export duties on nitrate. Until 1884 this arrangement proved particularly beneficial for its Iquique branch, since export duty payments were running at a rate of 500,000 pesos per month, of which the state was utilizing only 100,000 pesos.[30] The Banco de Valparaíso had the account for the Nitrate Railways Company until 1887, when the company placed its financial affairs in the hands of the Banco Nacional.[31]

Given such opportunities, the banks rapidly extended their credit facilities. By 1884 the Banco de Valparaíso and the Banco Nacional had between 4.5 million and 5 million pesos invested in nitrates. These figures probably represent only their regular accounts with *salitreros*, since by 1887 bill-discounting operations in Tarapacá were running at an annual rate of 25 million pesos.[32] This sudden expansion of the credit market was apparently due to the three banks' deposit operations in connection with the nitrate industry, since none of them increased their paid-in capital (a total of 10.7 million pesos) before 1887.

The overall performance of these institutions gives some idea of the profitability of their nitrate operations. With the Chilean economy at its nadir in 1878, the Banco Mobilario reported profits of 121,602 pesos and a dividend of 10 percent. But it earned 272,084 pesos in 1883 and issued a dividend of 24 percent. Both the Banco de Valparaíso and the Banco Nacional enjoyed similar renewals of prosperity as they regularly reported dividends ranging from 14 to 20 percent after 1880.[33]

With this profitable direct link to the nitrate industry and the general resurgence of the domestic economy, a marked improvement occurred in the national credit market. Interest on bank loans in central Chile, running as high as 11 and 12 percent since 1873, fell to 7 percent in 1880 and remained at approximately the same level for the rest of the decade. Credit extended by commercial banks in the form of loans, advances, and discounts grew from 59,961,000 pesos in 1881 to 102,312,000 pesos in 1888.[34] But as their operations in Tarapacá suggest, this growth occurred without any fundamental restructuring of the credit market. The banks in Tarapacá engaged in short-term discounting operations and provided credit secured by land mortages, much as they did in the Central Valley. Thus, the dawning of the Nitrate Age did little to alter "the elitist nature of the money market."[35] The expansion of the credit system turned to the advantage of the landed oligarchy, who enjoyed increased capital availability at reduced interest rates. The reason for this is evident from the ownership structure of two of the nitrate banks. The Banco Mobilario was virtually the private preserve of the Subercaseaux clan, a prominent estate-owning family. The list of the Banco Nacional's shareholders read like a who's who of the landed elite, led by Chile's greatest landowning family, the Ossa, who controlled 5 percent of its stock.[36] Whereas the effects of the nitrate region on the financial sector can be clearly established, the impact on mercantile interests is more difficult to determine.

The composition and operation of the Chilean merchant community has received scant attention from historians, since most of the nation's export trade, including nitrates, remained in the hands of foreign commercial houses.[37] Chilean interests appear to have centered around retailing of imported consumer goods and the coasting trade. Activity in both areas increased significantly because of the nitrate industry. Exports of nitrate made possible a sharp increase in imports, from 29,716,004 pesos in 1880 to 67,889,079 pesos in 1890. The value of the coasting trade skyrocketed from 34,109,477 pesos in 1878 to 103,792,745 pesos in 1890, with Tarapacá accounting for 20 percent of the total.[38]

In 1884 the newspaper *La Epoca* pointed to the close connection that now existed among Chile's agricultural, commercial, and financial sectors and her new province when it reported that:

Tarapacá, lacking agricultural products and in general all articles indispensable for subsistence, offers a very advantageous market to those who send them from further south, and even to the supply of foreign goods made in connection with the commerce of Valparaíso. The capital invested there has been supplied in large measure by institutions established in the country, and which have contributed mightily to the development of the industry.[39]

The observer could well have added the industrial sector to the list of the nitrate industry's beneficiaries.

Even before the nitrate zone began spurring economic growth, Chile's small industrial base had received stimuli both from the increase in import duties during the fiscal crisis of the 1870s and from the demand for manufactures for the war effort. During the first decade of the Nitrate Age, manufacturing was marked by improvements in production methods as well as by general expansion, as 846 new factories came into existence between 1880 and 1889. Although most of the nitrate industry's capital goods were still imported, it provided at least a limited market for such domestically produced durables as rail equipment and refining vats.[40] Probably more significant were the indirect encouragements to industrialization.

To the extent that it contributed to urbanization, the nitrate industry helped establish the preconditions for further industrialization. The increase in national income accelerated demand for domestic manufactures. In addition, nitrate exports produced the foreign exchange for purchase of raw materials and capital goods essential to industrial growth.[41] The linkages between nitrates and manufacturing suggest that at least in this area, nitrate wealth promoted an economic activity emphasizing innovation and productivity improvement that would require the transformation of Chilean society. But such was not the case.

If Chilean industrialization was a response to increased opportunities in the domestic economy, it occured despite structural impediments within that same system. Low productivity in agriculture served as a break on the industrialization process. The limits of the national credit market and the abysmal status of educational institutions were equally serious obstacles to industrial growth. Manufacturers were drawn primarily from the oligarchy that enjoyed access to the money market and from immigrants who possessed the necessary technical knowledge. Even then, industrial enterprises depended on foreign sources for additional capital as well as for technology.[42]

Machinery imported from Europe limited the labor demands of industry; European technicians and skilled workers precluded a radical restructuring of education; and European capital left the national credit market's orientation toward land mortgages intact. Thus, industrialization advanced through a direct partnership with European capitalism. This relationship overcame the seemingly inevitable conflict between industrialization and a traditional agrarian society. Emerging in this manner, industrial growth failed to become a focal point for challenges to the traditional institutions of Chilean society. The mining sector provided further evidence that significant structural changes did not take place.

Mining, once the principal buttress of the Chilean economy, did not share in the general postwar resurgence. The decline in world prices for copper and silver that began in the 1870s continued to hamper Chilean mining as the prices for both metals fell by 20 percent during the 1880s. Output did not fall as drastically as the trajectory of world prices would suggest. Copper production fluctuated until there was a drastic drop in 1889, while silver production actually increased. There were even periodic boomlets such as the one in 1882 when investors abandoned their bank stocks to invest in the companies working the Cachinal silver deposits. The decline in interest rates encouraged investment in mining companies and the abolition of mineral export duties reduced the cost at which Chilean products could be placed on the world market. In addition, a 22 percent depreciation of the peso's exchange rate between 1880 and 1890 had a favorable impact on production costs by holding down real wages of mineworkers.[43]

But elimination of export duties and currency depreciation, functioned as only temporary breaks on the decline of the mining sector. Unlike agriculture, mining still had to compete in the world market against more efficient producers. It was generally recognized that Chile had lost its dominant position in the world copper trade and that only massive infusions of capital and technology could rescue mining from continued disintegration. Although the number of Chilean joint-stock mining companies increased from 69 to 135 between 1880 and 1890, the average capital per company declined by more than 12 percent. The decreasing concentration of capital indicated that such momentous changes were not in the offing for the mining sector.[44] The downward slide continued until after the turn of the century when North American capital and machinery refurbished copper mining.

The industrial and mining sectors provide further evidence of the true form of Chile's nitrate prosperity. Unparalleled economic growth took place within the context of a traditional society. Obstacles to economic activities requiring capital concentration and innovation remained entrenched within Chilean society and could be overcome only with inputs of such factors from the capitalist centers of Europe and North America. The persistence of such barriers reflected the elite's need to preserve existing class relations. Meanwhile, the stock market bore daily testimony to the success of this experiment in grafting modern economic enterprises onto a premodern society.

The stock market played a major role in generating the nitrate and mining boom of the early 1870s but suffered a serious setback in the second half of the decade. Particularly hard hit were banks whose capital dropped by more than 10 percent and insurance companies that experienced a 37 percent decline in capital.[45] As late as 1880 the total authorized capital of Chilean joint-stock companies was down 10 million pesos from its high of 159,562,000 pesos in 1875.[46] Renewed prosperity brought on by the nitrate industry, and in particular the expansion of low-interest credit facilities, set off a new boom in corporate stock investment. Between 1880 and 1890, the number of joint-stock companies listed on the Valparaíso exchange increased from 144 to 304 while their total authorized capital rose from 150 million to 238 million pesos.[47]

Activity on the exchange followed the courses of the new nitrate wealth flowing through the Chilean economy. Investment capital gravitated to financial institutions, municipal service companies, and firms involved in commercial and shipping activities. As Tables 6 and 7 demonstrate these enterprises experienced a steady growth in stock values and paid healthy dividends in the years following the occupation of Tarapacá.

The companies listed in Tables 6 and 7 were among the consistent leaders on the stock exchange in terms of share prices and dividends. But, as the *South American Journal* noted in 1887, numerous other companies on the exchange were returning dividends of 4 to 9 percent. The journal also pointed out that these companies were the work of native capital and concluded that "Few countries can show such a list of remunerative enterprises."[48]

Of particular interest in the new stock-market boom is the composition of the investing public. The fragmentary records of corporation stockholders in this period indicate the predominance of the Valparaíso business community. But what is striking is the consistent appearance of distinguished landowning families such as the Ossa, Larraín, Echeverria, and Garcia-Huidobro.[49] It would be difficult to gauge the importance of such activities to the elite, but they indicate that direct support of a seigneurial life-style was not the sole use to which the elite put their new wealth. This involvement of the oligarchy was not new. It continued a pattern that emerged in previous decades and provided landowners access to higher profits than the paltry sums produced by most estates. What is significant is that the new link to the nitrate industry restored the viability of this crossover pattern that had been threatened by the economic decline of the late 1870s.

While these activities supplied diverse investment opportunities for the elite and ample evidence of prosperity, these improving economic conditions were accompanied by a growing awareness that the new wealth was inextricably linked to the fate of the nitrate industry. This awareness

Table 6
Share Prices of Leading Companies on the Chilean Stock Exchange, August 1877–86

Company	Paid-in Capital (Current Pesos) 1881	Share Prices — % of Paid-in Stock Value									
		1877	1878	1879	1880	1881	1882	1883	1884	1885	1886
Banks											
Valparaíso	5,125,000	70	56	92	135	140	158-1/2	136	151-1/2	136	150-1/2
Nacional	4,000,000	140	95	87	160	192	185-1/2	170	175	162	145-1/2
Agrícola	1,593,600	81	72	93	110	130	135-1/2	123	126	122	140
Mobiliario	1,125,000	80	80	—	—	—	145	150	150	160	—
Insurance Co.'s[a]											
American	200,000	150	150	155	250	335	330	310	300	296	268
Chilean	100,000	210	210	130	240	335	290	290	290	290	285
Union	100,000	55	55	48	20	92	112	110	109	110	98
Urban Service Co.'s											
Santiago Tramway	1,500,000	69	50	40	110	155	179	174	210-1/2	205	105
Valparaíso Tramway	600,000	165	150	125	155	186	218	245	223	232	205
Gas Consumers	150,000	85	97	200	190	265	250	235	245	305	340
Santiago Gas		n/a	n/a	n/a	n/a	n/a	130	125	137	167	157-1/2
Valparaíso Gas	400,000	—	160	—	166	185	151	150	174	200	220
Shipping & Commerce											
Valparaíso Dock	650,000	80	75	50	184	147	108	89	94	107	102
South American S.S.	2,191,500	12	35	88	140	192	137-1/2	115	103-1/2	124	146-1/2
Wood and Coal	500,000	40	35	30	79	90	145	135	124	132	126
Tugboat	300,000	77	72	80	150	202	170	130	137	140	126
Transandine Telegraph	500,000	85	90	125	130	156	162-1/2	114	118	123	137

Sources: *El Mercurio* (Valparaíso), August 1877–83; *SAJ*, 1 October 1887.

[a] Insurance company shares are listed in terms of share values (current pesos) rather than percentages.

Table 7
Dividends of Leading Companies on the Chilean Stock Exchange, 1875–77, 1879–81, 1884–86

Company	Paid-in Capital (Current Pesos) 1881	Dividends — % of Paid-in Stock Value								
		1875	1876	1877	1879	1880	1881	1884	1885	1886
Banks										
Valparaíso	5,125,000	8	8	9	11	14	8	16	17	18
Nacional	4,000,000	20	20	17	10	18	18	20	18	15
Agrícola	1,593,600	12	12	13	13	12	12	12	12	14
Mobilario	1,575,000	0	5	9	11	8	7	18	25	19
Insurance Co.'s										
American	200,000	10	25	25	46	50	35	40	35	30
Chilean	100,000	n/a	70	30	100	100	60	80	80	60
Union	100,000	10	24	28	22	26	24	20	26	14
Urban Service Co.'s										
Santiago	1,500,000	0	15	0	6	8	12	20	13-1/2	9
Valparaíso Tramway	600,000	5	9	4	14	16	15	21-1/2	17-1/2	16
Gas Consumers	150,000	8	22	24	18	30	34	40	40	40
Santiago Gas		n/a	n/a	n/a	n/a	0	0	13	15	16
Valparaíso Gas	400,000	n/a	n/a	n/a	0	22	20	20	20	20-1/2
Shipping and Commerce										
Valparaíso Dock	650,000	6	0	0	0	0	7	10	10	10-1/2
South American S.S.	2,191,500	0	0	0	6	18	16	13	18	19
Wood and Coal	500,000	n/a	6	0	3	9	8	16	22	14
Tugboat	300,000	6	12	12	17	18	32	6	13	14
Transandine Telegraph	500,000	0	12	14	12	12	20	16	19	19

Sources: *El Mercurio* (Valparaíso), 1875–82; *SAJ*, 1 October 1887

surfaced in the response of various sectors of the economy to crises in the industry.

Banking proved particularly sensitive to conditions in the nitrate sector. The failure of two of the smaller European *salitreros* in 1883 set off a panic in bank shares on the Valparaíso stock exchange. The shares of the Banco de Valparaíso and the Banco Nacional were especially hard hit by the brief plunge in stock values.[50] When the decline in nitrate prices threatened to ruin many *salitreros*, the Banco de Valparaíso became one of the principal advocates of a Combination. The bank also offered to make its credit facilities available to the Combination, an offer that was accepted.[51] The clearest indication of the sensitivity of Chilean financial interests to developments in Tarapacá came in an incident that occurred in the Valparaíso offices of the Gibbs house in March 1884. At that time it had been learned that a number of European powers were planning to bring pressure on the Chilean government because of the Peruvian bondholders' dissatisfaction with the terms of Chile's peace treaty with Peru. This raised the possibility of an increased export duty on Tarapacá in order to provide further compensation to the bondholders.[52] On March 3, one of the Gibbs partners sent the following report to the London office:

> On 1st inst. Don Melchor Concha y Toro called on the writer and said he wished to have some conversation about these matters—that he had no personal interest in them but felt alarmed at the turn things were taking—looking to the welfare of the country + the "comercio,"—that he feared the upshot might be that Tarapacá might be handicapped with some differential duty as compared with Antofagasta + the other non-Tarapacá production, and that he therefore wished to impress upon us the advisabalness of writing to you and pointing out this very probable result of the intervention for the chance of your being able to bring to bear any influence against it.
>
> Notwithstanding his disinterested and patriotic preamble, there is no doubt that Don Melchor has a strong personal interest in this matter, his and the Subercaseaux families being the principal owners of the "Banco Mobilario" which (like ourselves) would be most seriously prejudiced by any such difference of duties, the Bank having a large amount of capital employed in habilitating "Folsch + Martin" and Otto Hermann's oficinas. There can therefore be no doubt about the genuineness of Don Melchor's apprehensions.[53]

Like the banks, Chilean agricultural interests responded to conditions in the nitrate industry that might adversely affect their newfound prosperity. In 1883 their official publication, the *Boletín de la Sociedad Nacional de Agricultura,* extolled the advantages of the domestic over the export market.[54] Within five months of the initiation of the First Combination, however, conditions had changed dramatically. The labor force at the *oficinas* had shrunk by 33 percent, and the number of animals in use by 19 percent.[55] This sudden drop in the consumer market of Tarapacá was quickly answered by another article in the *Boletín*. Combining an unusual solicitude for the fate of the unemployed nitrate workers with the desire for a healthy agricultural sales climate, the author urged increased nitrate production at a lower price in order to reestablish full employment.[56]

After the occupation of Tarapacá, Chilean society enjoyed a new era of prosperity while preserving its premodern class relations. By their actions, those who composed the oligarchy clearly indicated an awareness of the close tie between these phenomena and the foreign-owned nitrate industry of Tarapacá. As a result, they could view foreign control of the industry with equanimity and even preceive the general interests of foreign *salitreros* as their own. The Chilean state reflected this passive and even approving attitude.

Chile's president, Domingo Santa María, viewed developments in Tarapacá with a sense of relief, inasmuch as the difficult international conflict over the nitrate certificates had been reduced to one that could be settled in the law courts. As for the inevitable decline in state revenues resulting from the foreign-dominated Combination, Santa María was equally unperturbed and stated that "What is essential is not to throw away like mad men in the street a wealth which only we possess."[57]

The relative contentment of government officials cannot be taken as a sign of ignorance about the growing power of foreign interests in Tarapacá. In September 1884 Gonzalo Bulnes, the intendant of Tarapacá, visited the port of Caleta Buena and made the following report to the minister of finance:

> Campbell Outram and Co. has created on that shore a veritable factory of its commercial house. Everything that one sees there belongs to it. With the exception of some workers' homes, the dwellings are constructed by it and commerce in all its spheres from merchandise to articles of subsistence belongs to the referred to house. Water itself is supplied by it to the fullest degree so that it could if it wished to abuse the incontestable monopoly of its capital, oblige any person or industry to leave Caleta Buena by denying it water and subsistence...
> The monopoly of the house of Campbell Outram and Co. has reached such lengths that no other type of money circulates in Caleta Buena but the scrip of the house itself which is only convertible into merchandise and only in its own stores.[58]

But Bulnes did not suggest any drastic measures to curtail the powers of this state within a state. He simply suggested the organization of a *gremio de jornaleros* (stevedores' union) to ensure efficient collection of the nitrate duty.[59]

Two years later, the new intendant of Tarapacá reported on the operations of the Banco de Valparaíso and summarized in glowing terms the relationship between Chilean capital and foreign *salitreros*:

In this manner it results, in the final analysis, that the native capital available in the south of Chile due to the customs of work and frugality, is a powerful auxiliary of the nitrate industry in the north, associating itself in diverse combinations with the active, intelligent and happy foreign entrepreneurs who opportunely send their huge profits to Europe in pounds sterling.[60]

Probably no single statement better summarizes the symbiotic relationship between the nitrate industry and Chile's domestic economy. Yet the relationship was not without its contradictory aspects.

As the response of agriculturists to the First Combination suggests, domestic interests favored steady increases in nitrate output as a stimulus to their growth. Indicative of the impact of production restrictions was the effect of the Combination on Chilean foreign trade. Owing to the reduction in nitrate exports, Chile's imports plummeted from 54,447,061 pesos in 1883 to 40,096,629 pesos in 1885.[61] The state was particularly sensitive to production levels, since its revenues depended on the amount of nitrate produced rather than on its price. Nitrate producers, on the other hand, considered output restrictions essential to the restoration of profitable operation. The divergence of interests over production levels, however, was transitory. More fundamental were the contradictions between the modern capitalist enterprises of the north and the traditional socioeconomic system supplying them with factors of production. But features of both the modern and traditional sides of the equation operated to overcome these difficulties.

During the 1880s agriculturists expressed renewed concern over problems of labor scarcity as migrations to the nitrate zone continued. This perception of scarcity reflected the need to maintain a significant surplus work force and static wage levels to meet the needs of a labor-intensive agrarian system. However, service tenantry did not require rigid retention of the entire rural populace in the countryside where nonwage methods of control still rooted an abundant labor force. Indeed, the wages of *inquilinos* remained stable, and hacendados rejected more exotic measures such as the use of indentured coolie laborers, indicating that competition for labor had not reached a point that would require agriculturists to restructure their methods of control.[62] Conversely, *salitreros* encountered labor supply problems resulting from the limits on the free wage labor market. In turn, relative labor scarcity facilitated the growing militancy of nitrate workers (see Appendix). Meanwhile, low productivity in agriculture contributed to increasing food prices, putting inflationary pressure on the *salitreros'* labor costs. The industry's increases in productivity served as a partial solution to these difficulties. The Chilean state offered assistance by legalizing the *ficha* (company scrip) system, enabling *salitreros* to lower real wages.[63] It also took steps to deal with challenges from the nitrate workers.

The government extended its system of *gremios de jornaleros* to all ports in Tarapacá. This not only improved the accuracy of nitrate duty collections but was also of critical importance to nitrate producers, since the dockworkers represented a small segment of the labor force capable of choking off nitrate exports.[64] Strikes occurred despite these efforts to create a docile work force in the ports, and government troops were brought in to load nitrate. When *oficina* workers struck in protest against the *ficha* system in 1890, the government deployed troops to restore order and allowed the *salitreros* to renege on promises of higher wages.[65] Through these policies the Chilean state sought to bridge the gap between the limited capabilities of the domestic economy and the modern capitalist enterprises in the northern provinces. State action on behalf of the nitrate industry was essential, since Chile's economic growth and preservation of its existing social structures now depended on the relationship with the nitrate industry.

The impressive quantitative economic changes wrought in the first decade of the Nitrate Age masked a lack of fundamental structural change. Government expenditures on transportation, education, and the bureaucracy; the resurgence of agriculture, finance, and commerce; as well as expansion of the industrial base resulted in large measure from linkages between the domestic economy and the nitrate industry. A coherent elite, an effective central government, and the economic infrastructure that had emerged before the war enabled Chile to derive substantial benefits from the nitrate industry. More important, these changes marked a qualitative transformation of Chile's relationship to European capitalism that facilitated a fundamental continuity in domestic social productive relationships.

A traditional society prospering from export sectors characterized by labor-intensive practices and low productivity, Chile had faced a decisive crisis in 1878. Its export markets were rapidly falling to more efficient producers, and—short of a radical and improbable restructuring of its society—economic retrogression appeared inevitable. Absorption of the nitrate industry preempted the seemingly inevitable decline. European capitalism, propelled by the desire to lower the cost of raw materials, created within Chile's national borders an industry capable of rapid increases in productivity. In essence, a fragment of the European capitalist economies had been transplanted into a traditional society. This marked a transition in an economy previously centered on domestically controlled labor-repressive, labor-intensive means of production linked to European capitalism by commercial relations. The base of continued economic growth was now a foreign-owned means of production characterized by productivity improvements. A truly symbiotic relationship emerged as the industry stimulated the domestic economy, which in turn supplied the capitalist enterprises with unskilled labor, operating capital, and foodstuffs.[66]

Rather than undermining the traditional structures of Chilean society, this direct implant of European capitalism revitalized the existing order. Symptomatic of this fact was the newfound prosperity of agriculture achieved within the existing premodern social productive relations of the countryside. Furthermore, state revenues derived from nitrate lowered transportation costs for the agrarian sector and opened new positions in the bureaucracy for the elite. So too, enhancement of the national money market placed vast new sums of capital in the hands of landowners for either conspicuous consumption or high-return investments. The existing political and economic infrastructure thus channeled wealth back into the domestic economy in such a way as to enhance rather than challenge the traditional society. The future potential of this new relationship is indicated by the courses of industry and mining. Expansion of the industrial base was achieved with infusions of European capital and technology that overcame but did not destroy domestic obstacles to industrialization. After the turn of the century, mining would be rescued from decline by similar inputs from the United States.

The failure of the nitrate industry to transform Chilean society was due to the premodern class relations that underpinned the position of the Chilean elite and the new relationship it established to European capitalism. The elite's social, political, and economic domination was a product of the social productive relationships of the countryside. The existence of the *inquilinaje* system rather than more extreme forms of forced labor such as slavery or serfdom and a readily accessible surplus rural population permitted a certain flexibility in the hacienda system. This flexibility is evidenced by Chile's economic progress before 1873 and the limited response of landowners to a continuing rural–urban migration. Yet the social, political, and economic power that emanated from the system made maintenance of its traditional structures essential. This limited the options the oligarchy could exercise in exploiting the new wealth. Acceptance and encouragement of foreign control of nitrates permitted the elite to enjoy renewed prosperity while avoiding the unacceptable alternatives of economic decline or of a restructuring of Chilean society that would undermine their position. In the nitrate industry, inputs of European fixed capital investment and technology as well as managerial and technical skills were overcoming the conflict between modern capitalist enterprises fueled by productivity increases and a traditional socioeconomic system. Foreign supply of such factors made it unnecessary to restructure Chilean society to produce these inputs domestically, while state taxation and market linkages regenerated the domestic economy. This new relationship was not without its contradictions, but the flexibility of the Chilean hacienda, the productivity of European capitalism, and the policies of the Chilean state helped surmount these problems. This process points to an important factor contributing to the lack of Chilean nitrate investments.

The failure of Chileans to invest directly in nitrates was a function not only of the increasing capital and technological requirements of the industry but also of the form of change in the domestic economy as well. The elite's need to preserve premodern agrarian structures led to acceptance of European control of the means of production that permitted economic growth but left obstacles to domestic development of the industry intact. The process of growth within the limits of a traditional society and the regeneration of the ruling elite that it facilitated become more apparent with a study of the career patterns of Chileans who had been involved in the nitrate boom of the 1870s.

[6]

Men Making Their Own History

"MEN make their own history, but they do not make it just as they please; they do not make it under circumstances chosen by themselves, but under circumstances found, given and transmitted from the past."[1] Marx's observation concerning the France of Louis Bonaparte might well be taken as an apt description of Chile's entrance into the Nitrate Age. Paced by the development of European nitrate production, it was an era of unprecedented growth and ostentatious wealth, truly Chile's Gilded Age. Its refurbished ports bursting with commerce, its principal cities ever increasing in population and enjoying modern urban services, a national government drawing on a seemingly inexhaustible treasury to link the country's regions ever more closely by rail and road and to expand its educational institutions, a rejuvenated stock market that captured in its feverish and diverse activities the scope of available economic opportunities, Chile appeared to be hurtling into that flawed but ever sought after utopia of modernity. Yet, as we have seen, these trappings of modern society cloaked an economic order whose essential structures remained highly traditional. Both the prosperous *fundos* (estates) of the Central Valley and the declining mining centers of the north symptomized the persistence of problems that had threatened the economy in the previous decade. Opportunities for profit proliferated, but within the limitations of a traditional society.

A labor supply constricted by non-wage-control mechanisms, acute maldistribution of national income, rising food prices owing to an inefficient agrarian sector, and the inadequacies of educational institutions typified the outcroppings of the past that constrained domestic economic development. Despite these constraints and the existence of dynamic entrepreneurial elements, a social consensus to sweep aside these barriers did not emerge. The strengthening of the elite's dual power bases, as incorporation of the nitrate industry prompted a resurgence of agriculture and state finances, provides part of the answer. But the absence of challenges was also the result of a more complex process at work within the upper strata of Chilean society. The career patterns of members of this group help define the parameters of that phenomenon.

Collective biography as a research tool can breathe life into the sometimes dry statistics that illuminate our understanding of social structures. Furthermore, it can often shed light upon subtle differences in complex social relationships. As has become apparent in studies of economic development, for example, the distinctions between traditional and modern or modernizing societies cannot be fully defined in terms of gross national product, income distribution, or other such statistical indicators. Decisions on the uses to which available resources will be put and the social structures conditioning those choices are of critical importance in making these distinctions. As employed here, collective biography is not based on the stringent methodologies of statistical analysis. No claim is made that the individuals whose careers we are about to examine are a representative sample of the upper echelon of Chilean society. Indeed, the basis of their selection, their involvement in the Chilean nitrate enterprises of the 1870s (see Table 8), is a deliberate effort to focus on individuals whose proven willingness to assume risks in new types of economic activities would place them among the more dynamic economic actors in the society. Their subsequent activities can be taken as at least an indication of the socioeconomic opportunities available to such elements and the general trend of the options they exercised. This approach, then, represents an effort to gain a clearer understanding of the forces shaping the Chilean response to the Nitrate Age.

Given our rudimentary knowledge of nineteenth-century Chilean society, it is difficult to place these individuals in precise positions within the social hierarchy. However, their wealth, which ranged from the modest estate of Antonio Costa valued at 116,000 pesos to the spectacular 9-million-peso fortune of Agustín Edwards, safely situated them in the upper-income group. A further broad distinction that can be made is their position within, or proximity to, the elite. While not the sole criteria for elite status, ownership of a rural estate was certainly the sine qua non for membership in the oligarchy. Other factors delineate a select group whose accomplishments propelled them beyond mere local or regional prominence. By the late 1850s, Santiago had achieved an unchallenged position as the center of national political power and the place of residence for the leading families who exercised that authority. Thus, a role in the national government and a home in the exclusive downtown section of the capital offered evidence of a position in the national elite. A further indicator that one had reached the pinnacle of Chilean society was membership in the prestigious social organization, the Club de la Unión founded in Santiago in 1864.[2] By these standards the ten individuals fall into three categories: Costa, Nechochea, Herrera, and Rosenberg, who never achieved such status; Ross, Délano, Edwards, and Oliva, who entered the oligarchy by the end of the century; and Subercaseaux and Concha y Toro, the scions of distinguished landowning families. It is within these rough definitions of economic and social status that their activities are traced, beginning with the group typified by Antonio Costa, who achieved only moderate wealth and little social distinction.

Within the thriving Valparaíso merchant community of the early 1870s, Antonio Costa distinguished himself as one of the most prolific investors in the Tarapacá nitrate enterprises. While the collapse of the Tarapacá firms put an end to his interests in nitrates, the flourishing of Chile's international trade after 1879 brought new life to his activities as a merchant. With his brother Emeterio he continued to operate the import-export firm of Costa Hermanos. The house business followed the growth trends in the Chilean economy, as it handled the import of consumer goods and supplied materials for the Valparaíso Tramway Company.[3] Increased demand for consumer goods and modernization of urban centers facilitated by nitrate wealth thus created new business for the Valparaíso merchant. As the holdings of his estate indicate, the resurgence of the national credit market and urban growth created investment opportunities for Costa, with the bulk of his capital flowing into financial institutions, urban service companies, and urban property (see Table 9). His activities suggest that it was quite possible to pursue a variety of undertakings that were unimpeded by the discontinuities within the domestic economy. The career of José María Nechochea offers further evidence of this fact.

Necochea had taken a far more direct role in the Tarapacá companies, serving as a stockholder and manager of two of them.[4] But like Costa his interest in nitrates ended with the collapse of those firms. During the 1880s Necochea derived his income from a variety of sources—working as an importing agent, dabbling in the purchase and sale of urban real estate, and serving as a government functionary.[5]

Table 8
Participation of Ten Chilean Entrepreneurs in Chilean Nitrate Companies

Name	Company	Number of Shares	Positions Held
Concha y Toro, Melchor	San Carlos	10	director
Costa, Antonio María	Negreiros	10[b]	n/a
	Negreiros	5	n/a
	Pisagua[a]	n/a	n/a
	Sacramento	50[b]	n/a
	Valparaíso	40	n/a
	Chilena	5	n/a
Délano, Eduardo	Sacramento	n/a	business agent of A. Edwards y Cía. in its Tarapacá nitrate operations, later owner of the *Oficina* Sacramento.
Edwards Ross, Agustín	Antofagasta	17[c]	director
Herrera, Oscar	Valparaíso	115	co-founder, secretary, director
Necochea, José M.	Pisagua[a]	n/a	general manager
	San Carlos	n/a	manager
Oliva, Daniel	4 *Oficinas*	n/a	founder and sole owner
Rosenberg, Gustavo	Solferino	22	n/a
	California	4	director
Ross, Jorge	Valparaíso	35	director
Subercaseaux, Francisco	San Carlos	30	co-founder

Sources: *Memorias* and *Estatutos* of Chilean nitrate companies, 1870–79; *NV*, 1870–78, Hernández, *Salitre*, pp. 84–86.
[a] The records of the Compañía Pisagua did not list the number of shares held by each stockholder.
[b] Signifies jointly held stock.
[c] The estate of Edwards's late father, co-founder of the company, held 1,049 shares in the firm.

Table 9
The Estate of Antonio María Costa, 1888

Holdings in Joint-Stock Companies:	
Banks	34,125.00 pesos
Insurance Companies	1,000.00
Urban Service Companies	9,165.00
Railroads	2,500.00
Mines	4,120.00
Miscellaneous	750.00
Urban Properties in Valparaíso	41,742.85
Personal Credits and Effects	19,693.81
Costa Hermanos in Liquidation	3,000.00
Total:	116,096.66 pesos

Source: NV, vol. 282, 19 June 1888, fs. 902–6.

The Valparaíso businessman thus found opportunities, though apparently limited ones, in the expansion of international trade, urban growth, and the development of the bureaucracy. His estate does not provide a cash valuation of his holdings, but the limited number of his stocks, his paltry savings, and the description of his properties as small indicate an individual of modest means (see Table 10). In view of his limited capital, it is interesting to note that Necochea's most persistent investment interests were in mining. In addition to the copper mining shares listed in the inventory, he was a stockholder in, and director of, an ill-starred mining company and invested capital in several small silver mines.[6] These efforts suggest reasons for the continued attractiveness of the troubled mining sector for individuals such as Necochea. It constituted an activity requiring only a limited capital investment, and it held at least the hope of spectacular returns. As in the case of Costa, there was little in the structure of these endeavors to prompt challenges to the existing system.

Neither Costa nor Necochea could match the third member of this group, Oscar Herrera, in diversity of interests. Herrera had been instrumental in the founding of the Compañía Valparaíso and played an active role in its affairs.[7] Subsequently he devoted his attentions to his mercantile establishment in Valparaíso that specialized in the importation of consumer items such as American-made watches.[8] By 1882 he held 15,000 pesos in shares of the Banco de Valparaíso and 5,750 pesos of stock in a tugboat company.[9] He later increased his involvement in financial institutions by purchasing 5,000 pesos of stock in the Valparaíso Insurance Company and helped found the insurance firm La Protecta, in which he served as vice

Table 10
The Estate of José María Necochea, 1890

House — Valparaíso	
Urban Property — Valparaíso	
Several Small Properties — Quillota	
20 Shares	— San Felipe Railroad
2 Shares	— Curico Gas Company
7 Shares	— Chuquicumata Mining Company
16 Shares	— Copper Mine Mañanita
8 Shares	— Copper Mine María Teresa
1,000 Pesos	— Public Employees' Savings Bank

Source: NV, vol. 301, 1 January 1890, fs. 671–77.

president.[10] With Valparaíso real estate values rising, Herrera bought a tract in the province for 8,000 pesos, subdivided it, and sold the property for 11,530 pesos.[11] But his activities extended beyond these fairly conventional forms of enterprise.

The quickening pace of industrialization attracted Herrera's attention. In 1890 he held 5,000 pesos in shares of the Fábrica Nacional de Cerveza and became one of its directors. It is noteworthy that the success of the beer industry at this time was due to the wholesale importation of European technology.[12] Herrera also promoted a monorail transport system for nitrates invented by his brother Demofilo. Gibbs and Campbell both displayed serious interest in the device, providing the capital for its testing and development, and Campbell made plans to employ it between its *oficina* Agua Santa and the port of Caleta Buena. However, the endeavor was eventually frustrated by the Nitrate Railways Company, which enjoyed a monopoly on rail transport in Tarapacá.[13] Herrera's career demonstrates not only the continuing opportunities in such areas as luxury-good imports but also the important role of foreign technology and capital in launching nontraditional endeavors.

The investments of Gustavo Rosenberg help distinguish the dividing line between the successful businessman and those who moved beyond that category of nouveau riche into the elite. After serving as a director of one of the most durable of the Chilean nitrate companies, Rosenberg became involved in a variety of joint-stock enterprises. His investments followed a common pattern, as he poured more than 45,000 pesos into financial institutions and urban service companies between 1882 and 1883.

He continued to pursue this course of action in 1887 with a 12,000-peso investment in the Compañía de Gas de Concepción.[14] His access to the national credit market was evidenced by bank loans totaling 61,500 pesos, which he obtained in 1883.[15] This access reflects the fact that his interests had already turned to the agrarian sector.

The Chilean government's final suppression of Indian resistance on the Araucanian frontier in 1883 opened the virgin lands of that region to rapid development. By that time Rosenberg owned two *fundos* in the area, and in 1885 he purchased two additional properties in a state auction.[16] Rosenberg had apparently secured one essential element toward elite status. Yet such a position did not prove to be his goal. The new estates of the south were themselves valuable investments. The frontier represented one of the few areas where agricultural production for export was actually expanding. And Rosenberg treated his estates as business enterprises. In 1882 he signed an agreement with one Juan Urrutia. Urrutia agreed to manage the estates for Rosenberg, who provided 36,000 pesos for their improvement, with the two men sharing profits from the enterprise.[17] Rosenberg treated his agrarian interests as simply another form of business investment. Even should they fail to return a profit, they could serve as an excellent form of collateral for bank loans. The absence of his name from the biographical compendiums that chronicle the activities of the Chilean elite further indicates that Rosenberg's *fundos* did not represent an effort to establish himself among the traditional hacendado class. As Rosenberg's career demonstrates, elite status was not defined simply by ownership of a *fundo*. Preferably such landholdings would be in the Central Valley and would be complemented by appropriate familial and political connections. For whatever reasons, Rosenberg failed to establish himself within that matrix of landed wealth and sociopolitical relationships.

In terms of social mobility, none of these individuals altered his status significantly. Their continued residence in the port city and their absence from the institutions of national political and social power in the capital indicate that they remained among the less prominent figures in the Valparaíso business community. For someone such as José María Necochea, exalted social status clearly lay beyond his financial means. For Rosenberg, it may have constituted a conscious choice to forgo social aspirations that would make him a target of ethnic prejudice. In any case, these entrepreneurs found little reason to challenge the limitations of a domestic economy that offered substantial opportunities in conventional endeavors such as finance and land speculation. The more venturesome, such as Oscar Herrera, could turn to foreign capital and technology to enter areas in which the domestic supply of such factors was inadequate. Nor were these individuals bereft of connections that could work to their benefit. Necochea was sufficiently well acquainted with the millionaire miner and landowner Federico Varela to obtain a personal loan from him; and Herrera became a business associate of the prominent banker Agustín Edwards.[18] As we shall see in tracing the careers of Edwards and his relatives, they were members of the business community who scaled the heights of the social promontory.

Jorge Ross, a director of the Valparaíso Nitrate Company, was a figure of some importance in the port of Valparaíso. A member of a successful merchant family, Ross in 1867 became a senior partner in A. Edwards y Compañía established by his cousin Agustín Edwards Ossandón. As manager of the firm's banking operations, his share in its profits in 1885 alone amounted to 45,000 pesos.[19] The Edwards bank was his principal concern, but he also had extensive outside business interests. In the early 1880s his holdings included 5,000 pesos in the Compañía Comercial de Remolcadores (a tugboat company), 5,400 pesos in the Santiago Tramway Company, and 2,000 pesos in the Valparaíso Insurance Company.[20] By 1887 Ross had become a director of a coal importing company and had invested 4,000 pesos in the Concepción Gas Company. He also served as the director of a municipal water enterprise in Valparaíso, and when the firm failed he assumed personal responsibility for its 70,000 pesos in debts.[21] This interest in urban services complemented his purchases of 154,000 pesos in urban properties between 1882 and 1886.[22] The dizzying array of his activities and the clear indications of substantial wealth leave little doubt of the prosperity Ross enjoyed in the first decade of the Nitrate Age. But, unlike the preceding cases, he successfully linked this wealth with the requisites for elite status.

In addition to a *fundo* in Coquimbo, Ross owned an estate in the province of Valparaíso, valued at 60,000 pesos in 1896. His home was located among the cluster of mansions that bordered the Plaza de Armas in the capital, and he was a member of the Club de la Unión. In terms of family ties, his wife bore the distinguished name of Carrera, and his daughter married the Chilean millionaire Arturo Lopez Pérez.[23] His son Gustavo became one of the most prominent and most reactionary politicians in twentieth-century Chile.[24] Another member of the Edwards clan who enjoyed a similar rise to social as well as economic prominence was Eduardo Délano.

Délano's involvement in the nitrate boom of the 1870s stemmed from his position as manager of the Edwards family's nitrate investments in Tarapacá. After the occupation of the province, Délano, in partnership with several other Santiago businessmen, purchased the Lagunas nitrate grounds for 115,000 pesos. Frustrated in his efforts to obtain a government railway concession for the property by the intervention of the Nitrate Railways Company, Délano sold Lagunas to John Thomas North in 1888 for 792,000 pesos. His share of the proceeds from this transaction was approximately 400,000 pesos.[25]

Other areas of the economy attracted his attention, including the nation's two largest banks in which he invested 35,000 pesos, the South American Steamship Company, and an ore refinery into which he poured a total of 20,000 pesos.[26] As was the case with Oscar Herrera, Délano became involved in the expansion of Chile's industrial base, investing in a brewery and a sugar refinery. The refinery, the Compañía de Refinería de Azucar de Viña del Mar, was the largest such establishment in Chile. Contributing to its success were the work of its founder, Julio Bernstein a German immigrant, and credit provided by foreign trading houses.[27] Activities in industry, however, did not create a barrier to elite status.

Much of the profit from the nitrate transaction with North appears to have been invested in the agrarian sector. In 1890 Délano purchased two wheat *fundos* and a mill for 166,000 pesos. By 1896 he owned three estates with a total value of 542,800 pesos.[28] His successful entrance into the upper echelons of society is further evidenced by his political career. Before his death in 1921, Délano served three terms in the Chamber of Deputies, became minister of industry under his close friend, President Pedro Montt (1906-10), and held a position on the State Railway Council.[29] But even Délano's accomplishments paled in comparison with those of the leader of the clan.

Agustín Edwards Ross was by far the most distinguished of these newcomers. Building on the fortune acquired in banking and commerce by his father, Agustín Edwards Ossandón, he became the wealthiest Chilean of his day. In addition to his interests in A. Edwards y Compañía, Don Agustín involved himself in almost every imaginable line of economic endeavor. Public utilities, mining, railroads, manufacturing, and publishing were counted among his investments.[30] Of his diverse undertakings, one of particular interest to this study is the Compañía de Salitres y Ferrocarril de Antofagasta, which his father had founded in partnership with Antony Gibbs and Sons. From its establishment in 1868 the firm relied on Gibbs for technology as well as for experienced technical and management personnel. Marketing facilities were initially provided by Gibbs, and subsequently by the French shipping magnate Antoine Bordes. Upon taking control of the firm in 1878, Don Agustín displayed an antagonistic attitude toward Gibbs, but the company continued to rely on the British merchant house and, later, on other foreign sources for essential factors of production. Through these direct links with foreign capital the Edwards family overcame the domestic economy's barriers to an enterprise requiring rapid increases in productivity.[31] As we have seen, the case was not unique, since the industrial sector depended on foreign technology, skills, and capital. Where opportunity beckoned in undertakings whose requirements conflicted with the prevailing socioeconomic structures, the contradictions were overcome through direct partnerships with European capitalists. Agustín Edwards's career epitomized the resolution of the seemingly inevitable conflict between the modern and the traditional.

At the time of his death in 1898, Edwards owned four large haciendas. Although up-to-date technology and techniques were employed on the estates, these properties were nonetheless stepping-stones to social prominence. The Edwardses' Santiago home, valued at 450,000 pesos, made them neighbors of such noted families as the Subercaseaux, while Don Agustín enjoyed the company of other leading members of Chilean society at the Club de la Unión. Edwards's six daughters married into distinguished Chilean families, and he himself served in the Senate and became minister of finance under Balmaceda.[32] The rise of Edwards and other nouveau riche elements into the elite was not due simply to some irresistible attraction of the land and of the life-style of a hacendado. There were practical rewards in assuming the role of a landed oligarch illustrated in the success of the Edwards family.

Landownership served as the most effective means of obtaining capital for investment in such areas as manufacturing. In terms of the political power it conveyed, the Chilean government's defense of the Antofagasta Company from Bolivian taxation was encouraged by the presence of members of the elite among the company's shareholders. They included twelve congressmen and two cabinet members, many of whom were close political associates of the Edwards family. Such influence could not prevent Chilean taxation of the company, but in 1884 it helped secure a badly needed government railway concession. From his own position in the Senate, Don Agustín urged duty-free importation of machinery and other products for industry. It was a matter of no small concern to Edwards, who had recently established an earthenware factory and was involved in a series of other industrial undertakings.[33] This linking of the social and political power of the countryside with the economic power of urban enterprises was not confined solely to the Edwards family.

Daniel Oliva first became involved in nitrate production in Tarapacá in

1859. Forced out of the province by the Peruvian expropriation, he shifted his base of operations to Taltal. Perhaps the proverbial exception that proves the rule, Oliva succeeded in a region where other Chileans failed, and apparently without a direct partnership with European capitalists. He operated four *oficinas* and acquired a fortune of 400,000 to 500,000 pesos by 1888.[34] When nitrate properties became the rage on the London stock exchange, Oliva sold his principal *oficinas* to the Lautaro Nitrate Company in 1889 for 1,785,000 pesos.[35]

Despite decades of dedication to the nitrate industry, Oliva's commitments now gravitated toward the countryside. His estate, El Sauce, made him one of the most important landowners in central Chile. However, the former *salitrero* did not slip into the role of absentee landlord of a neglected estate. His reputation as a progressive agriculturist was epitomized by the electrical power plant he built at El Sauce to provide power to the hacienda and surrounding areas. Yet there is every indication that Oliva enmeshed himself in the best society. He served for more than a decade as a deputy and then a senator while also fulfilling the demands of noblesse oblige with notable acts of philanthropy. After his death, his widow established herself in Santiago, that irresistible magnet for Chile's ruling class, leaving two of her sons to manage El Sauce.[36] Included in the privileged social circle that Señora Oliva joined were the Subercaseaux and Concha y Toro families.

For Francisco Subercaseaux and Melchor Concha y Toro, scrambling for status was unnecessary, since they already stood at the pinnacle of the social pyramid. Both were descendants of prominent families, the Subercaseaux being relative latecomers, having made their mining fortune in the early nineteenth century. The lives of the two men were intimately linked, not only as brothers-in-law, but also as partners in a number of businesses. After 1879 nitrate wealth flowed into their hands through the Banco Mobilario as it provided operating credit to the *salitreros* of Tarapacá. Subercaseaux's investment in the bank grew with its profit margin. In 1877 Don Francisco held 12 percent of Mobilario's stock. Between 1881 and 1883 he purchased additional shares, increasing the value of his holdings to 462,275 pesos, or 29 percent of the bank's paid-in capital. His investments in financial institutions also included 200,000 pesos in bonds of the Banco Garantizador de Valores.[37] As his wealth increased, Subercaseaux joined his brother-in-law in exploiting the opportunities opening up on the Araucanian frontier. They formed a company in 1887 that committed 60,000 pesos to develop 3,295 hectares of southern land they had purchased at a government auction for 128,976 pesos.[38] But much of Don Francisco's wealth remained and continued to flow into the estates of the Central Valley. In

1884 his share of the family *fundo* Pirque was valued at 200,000 pesos. Between 1881 and 1885 he acquired four additional estates at a total cost of 534,484 pesos. He then leased these properties, and by 1890 his annual rent from the *fundos* totaled 67,000 pesos.[39] Subercaseaux thus channeled wealth earned in financing nitrate production into the traditional countryside. From this base he pursued the classic life-style of the Chilean hacendado, enjoying the pleasures of a 300,000-peso home in Santiago, membership in the Club de la Unión, and twenty-six sojourns to Paris in the space of twenty years. So frequent and prolonged were his absences that Melchor Concha y Toro acted as his agent in most of his business dealings.[40]

Concha y Toro proved less susceptible than his brother-in-law to the enticements of the Old World. In addition to his holdings in the Banco Mobilario, he was one of the founders and the director of the Banco Garantizador de Valores.[41] The enterprise to which he devoted the greatest part of his energies was the Compañía de Huanchaca de Bolivia, a silver mining concern. He had been one of its directors since its incorporation in 1873 and now became its president. In 1884 the corporation joined with the Antofagasta Company to extend the rail line that linked Antofagasta's *oficinas* to the coast. The extension to the Bolivian border promised cheaper transportation costs for Huanchaca, and when difficulties arose between the two firms, Huanchaca assumed sole ownership of the line. However, the cost of the extension and upgrading of its own production facilities proved to be beyond the firm's capacity. In 1888 Concha y Toro traveled to London and incorporated the Antofagasta and Bolivia Railway Company on the stock exchange. This firm then purchased the rail line from Huanchaca and leased the rail facilities to the silver mining company.[42] Reminiscent of Edwards's experience with Antofagasta, Concha y Toro overcame obstacles to his business endeavor with direct links to British capitalist institutions. And in a further parallel, Don Melchor's activities spanned the modern and traditional sectors of the economy.

Some of his agricultural investments reflected those innovations that characterized the haciendas of Oliva and Edwards. In addition to his agribusiness with Subercaseaux, both families were involved in the Viña Concha y Toro, a product of diversification in agriculture in response to a growing domestic consumer market.[43] But his agrarian holdings also included the more traditional type, specifically one *fundo* that he rented for 3,500 pesos per year and a second valued at 109,500 pesos in 1895.[44] His political career offered further evidence of his position in the oligarchy. From 1864 until 1886 he served as a member of the Chamber of Deputies, a tenure that included a brief term as minister of finance and a series of leadership posts in the Chamber. This was followed by a five-year term as a senator from Santiago before his death in 1898. From his strategic political position, Concha y Toro opposed the occupation of Antofagasta that threatened his Bolivian interests, promoted the uniform nitrate duty that favored Tarapacá where the Banco Mobilario had committed its funds, and supported Antofagasta's railway concession to the benefit of the Huanchaca Company.[45]

The careers of Subercaseaux and Concha y Toro indicate that if the wealth of the Nitrate Age offered opportunities for advancement to the nouveau riche, it also allowed more ambitious members of the oligarchy to expand and strengthen their positions. The Nitrate Age provided social mobility to a select group of newcomers, but it did not condemn the oligarchy to wither away en masse before their advance.

The very diversity of the endeavors of these ten individuals makes impossible a simple summary and conclusion about their significance. They do suggest, however, certain trends that were already emerging in the first decade of the Nitrate Age. Economic growth tended to focus in areas such as finance, commerce, urban development, and agriculture where market linkages and state revenues funneled wealth from the nitrate industry. The resurgence of agriculture, financial institutions, and state revenues strengthened the traditional elite that controlled them. Nor did expansion in any of these areas represent an inherent challenge to the existing order. This growth also made it possible for the most successful elements of the business community to buy their way into the elite with great estates and the arrangement of suitable marriages. This represented a practical decision, for as the activities of Edwards and Concha y Toro demonstrate, political power that emanated from the countryside could be a valuable lever in furthering economic interests. Newcomers like Edwards and Oliva brought innovation to the agrarian sector, but their model estates remained exceptions to the generally backward condition of Chilean agriculture. The eventual immersion of their offspring in the value system of the elite is typified by Gustavo Ross, whose political supporters described him as a twentieth-century Portales.[46] At the same time, the more farsighted members of the oligarchy involved themselves in profitable urban ventures, aware that a rural estate, for all its sociopolitical significance, was more likely to consume than increase their fortunes. The merging of the elite with the newly rich is symbolized by their sharing of the trappings of power including estate ownership, national political office, residence in Santiago's most exclusive neighborhood, and membership in the Club de la Unión. In rejuvenating much of the domestic economic order, the nitrate industry reinforced and augmented the traditional elite that it served.

Although its full significance would not become apparent until the twentieth century, there was a complementary process acting to overcome potential contradictions elsewhere in the system. Edwards's experience with Antofagasta, Concha y Toro's involvement in Huanchaca, and numerous investments in industry such as those of Herrera and Délano indicate that a variety of enterprises required capital, technology, and skills that were not readily available within Chilean society. The talent of European immigrants, direct ties to European trading houses, and the London capital market successfully overcame those barriers, precluding challenges to the traditional order.

These collective biographies point to the complex network of forces that helped preserve the traditional order during the Nitrate Age. The resurgence of existing areas of the economy benefited both the oligarchy and those who aspired to it. New economic endeavors whose requirements might conflict with the archaic institutions of Chilean society were pursued with the aid of foreign capital, preempting demands for far-reaching reforms. Since estate ownership provided access to the money market and political power, entrance into the oligarchy became not only a matter of social status but a means of promoting economic interests. Links to the European capitalist system in the nitrate industry and direct partnerships with it in the industrial sector made it possible for the Chilean economy to assume aspects of modernity while the ruling elite was preserved and enhanced.

Thus, Chile continued to be ruled by an elite whose political power and social values flowed from the traditional countryside while their economic interests encompassed both modern and traditional sectors without arousing fundamental conflicts. It proved a highly stable socioeconomic order but one unlikely to produce a social consensus for the progressive reforms that precede the emergence of mature capitalism.[47]

These developments did not constitute a simple extension of the adaptive process that characterized Chilean society prior to 1879. The primary motive force was no longer domestically controlled means of production but the modern European nitrate enterprises incorporated within Chile's national boundaries. The traditional order could not escape entirely unchanged in this new relationship. An emerging militant working class had already made its presence known in the northern desert, although its full impact would not be felt until the turn of the century. Of more immediate concern to the elite were the changes the nitrate industry underwent between 1886 and 1890 and, most important, the new role thrust upon the state by the Nitrate Age. These changes would converge in 1891 and shake the Chilean state to its foundations.

[7]

The Emergence of Monopoly Capitalism

IN his annual message to congress on June 1, 1889, President José Manuel Balmaceda delivered a stern warning against the development of a foreign nitrate monopoly.¹ Balmaceda's concern was certainly justified, for during his administration foreign domination of the nitrate region reached unprecedented proportions, with British companies alone accounting for 69 percent of the industry's output (see Appendix). But the significance of this growth in European control extends far beyond its quantitative dimensions.

With the means of production for its most important resource in the hands of Europeans, Chile was more directly affected than ever before by changes in the capitalist center. As Europe plunged into the era of monopoly capitalism, this transition in capitalist organization rapidly penetrated the nitrate region, reshaping the ownership structure of the industry and eventually having a significant impact on the Chilean political process. The collapse of the First Combination of nitrate producers in 1886 set in motion these momentous changes.

As we have seen, the effects of competitive capitalism upon the nitrate industry reached their apogee after the occupation of Tarapacá. Capital concentration made possible a tremendous upsurge in productive capacity and a sudden acceleration of productivity improvement. These developments contributed to both an oversupply crisis and to its temporary resolution in the First Combination. Continued increases in productive capacity created growing pressure from the lowest-cost producers to break the Combination, leading to its collapse in December 1886. The end of the

Combination, however, did not cause a sudden plunge in nitrate prices or profits. During the latter stages of the Combination, producers had enjoyed record earnings. The resumption of competition led to some reduction in price, but producers still earned healthy profits.[2]

The reasons for the persistence of prosperity were fairly simple. Increases in productive capacity and further implementation of the Shanks refining system had improved productivity during the Combination's life span (see Appendix). At the same time, a sharp rise in world consumption prevented a drastic drop in nitrate prices (see Table 11). The profitable condition of the industry now made it an inviting prospect for investors, and by 1887 developments in England provided a ready source of capital for such investments.

As Marx noted, capitalism's drive to increase the scale of production pushed accumulation beyond the stage of capital concentration by individual industrial capitalists to what he termed centralization of capital. Centralization could be achieved by "the amalgamation of a number of capitals, which already exist or are in process of formation... by... forming stock companies."[3] And as Marx suggested, this process is a step on the road to monopoly capitalism.

The model for Marx's theory was most certainly the British economy, where as early as the 1840s the London stock exchange took an increasingly important role in supplying fixed capital to industry.[4] This growth in stock-market activity assumed phenomenal proportions in the 1880s. In that decade, the number of companies registered with the exchange grew from 1,074 to 2,542, and their total authorized capital nearly tripled from £83,932,351 to £222,253,402.[5] At the same time, a deepening of the investment community was in progress. While the wealthy dominated stock investments,

> Improvements in communications, the development of Joint-Stock finance, compulsory education, the Married Woman's Property Act, and a rising standard of living were bringing about a striking change in the personnel of the investing public. Retailers, professional men, skilled workers and women were all being attracted to the Stock Exchange.[6]

In the early 1870s British investors had subscribed heavily to the bond issues of foreign governments, including the Latin American states, only to see their dividends evaporate when a number of these countries defaulted on the loans.[7] During the 1880s, however, the public showed a renewed enthusiasm for foreign investment, and Latin American economic enterprises constituted an important part of their portfolios. The total nominal

Table 11
The World Nitrate Market (in Tons), 1886–90

Year	Exports	Stocks[a]	Consumption	Average Price (£)
1886	444,000	282,000	323,000	10/5
1887	700,000	390,000	443,000	9/10
1888	770,000	429,000	598,000	9/10
1889	905,000	580,000	762,000	10/0
1890	1,028,000	710,000	885,000	8/14/6

Sources: Data on exports, stocks, and consumption for the period 1886–89 are from Chile, Ministerio de Hacienda, *Memoria de 1890*, pp. xl, lxxx–lxxxi. Data on exports, stocks, and consumption for the year 1890 are from *SAJ*, 17 January 1891. Information on prices for the period 1886–89 is from the *Statist* (London), 14 December 1889. Information on prices for the year 1890 is from *SAJ*, January 1890 to December 1890.
[a] Includes nitrate stored in consuming countries and in shipment from Chile.

value of British investment in Latin America grew from approximately £172 million in 1879 to £422 million by the end of 1890.[8] This sudden surge of capital into Latin America did not result from some capricious turn of mind by British investors. As the capitalist system achieved ever greater increases in labor productivity, the cost of raw materials, imported from peripheral areas where labor-intensive methods of production were predominant, came to represent a disproportionate share of the costs of production. British capital now poured into such areas as Latin America in an effort to counteract this trend.[9] This is apparent in the altered composition of these investments. In 1875 government securities constituted 74 percent of British investments in Latin America. In the following decade, £70 million was invested in such bonds, and £180 million flowed into economic enterprises.[10] Most of these enterprises were railways, which served to cheapen the cost of shipping raw materials to Europe. In nitrates, European merchant capitalists had spearheaded this drive, and their efforts paved the way for the direct entrance of British monopoly capital into the production process. These dynamics of capitalist development in Europe combined with his experience in nitrates made John Thomas North one of the most prominent stock promoters of the nineteenth century.

Colonel North, as he was soon to be known to an admiring public, had earned a reputation as a successful nitrate entrepreneur even before the collapse of the First Combination. The Liverpool Nitrate Company, launched in 1883, and the Colorado Nitrate Company, founded in 1885, had been highly remunerative enterprises. In its first three years, Liverpool paid out annual dividends of 26, 20, and 40 percent, respectively. The Colorado enterprise produced initial dividends of 8 and 15 percent. Given this impressive record, North found it relatively easy to organize yet another firm, the Primitiva Nitrate Company, in 1886.[11]

To further his career as a promoter of nitrate enterprises, North not only relied on this unquestioned business success but also went to great lengths to create a name for himself in British society. At his lavish estate at Avery Hill in Kent, North entertained the cream of British society. Randolph Churchill and Nathan Rothschild were only two of the many prominent figures whom North counted as his friends. His free-spending ways also earned him a reputation as a sportsman and a philanthropist as well as the honorary title of Colonel.[12]

Enjoying a reputation as a successful nitrate entrepreneur and a man of some social prominence, North, with the able assistance of Robert Harvey, took advantage of the favorable conditions in Tarapacá and London to launch a series of new nitrate companies. In 1888 the San Jorge and San Pablo Nitrate companies were formed, and 1889 witnessed the appearance of the San Donato and Paccha-Jazpampa enterprises.[13]

North earned spectacular promoter's profits from these transactions. The certificates for the *oficina* Ramírez, which North and Harvey sold to the Liverpool Company for £50,000, had cost them £5,000, a profit of 1000 percent.[14] The Paccha and Jazpampa *oficinas* had been operated by North through two private firms with a total capitalization of £64,530. When he sold the two establishments to the company in London, the price was £340,000.[15] The demand for the shares of North's companies proved equally impressive (see Table 12).

The promotion of nitrate companies appeared to be a one-man operation. In October 1888 the *South American Journal* noted that:

> It may be witchcraft, or it may be luck, or it may be and probably is, an unusually keen perception of facts, combined with an extraordinary astuteness, but it is a fact that everything that Colonel North touches seems to turn into gold and all his ventures prosper. The consequence is that his motions are observed, and many people follow his lead without hesitation.[16]

Soon even North's financial wizardry could not keep pace with the demand for new nitrate investments, and a host of other companies joined his own on the stock exchange. By the end of 1889 the total authorized capital of nitrate production firms registered with the exchange stood at £4,820,000 (see Table 13).

While North enjoyed most of the publicity in the London nitrate boom, other *salitreros* also secured enormous profits from these operations. In 1883

Table 12
Share Prices of North's Nitrate Companies, 1889

Company	Fully Paid Share (£)	Highest Price 1889 (£)
Liverpool	5	30-1/2
Colorado	5	12
Primitiva	5	38-1/2
San Jorge	5	13-1/4
San Pablo	5	15-1/2
San Donato	5	8-1/8
Paccha-Jazpampa	5	n/a

Source: SAJ, 17 January 1891.

Henry B. James and George Inglis valued their 60 percent share in the *oficina* Puntunchara and their 45 percent share in the *oficina* Tres Marías at less than £61,674. Yet when they sold Puntunchara to the London Nitrate Company in 1887, they received £137,500, despite depreciation of the grounds and the fact that the productive capacity of the *oficina* had not been

Table 13
Nitrate Production Companies Registered with the London Stock Exchange as of 12 December 1889

Company	Date Established	Authorized Capital (£)
Liverpool	3 February 1883	150,000
Colorado	13 June 1885	160,000
Primitiva	8 July 1886	200,000
London	3 August 1887	160,000
San Pablo	8 August 1888	160,000
Taltal	10 August 1888	85,000
Santa Luisa	2 November 1888	250,000
San Jorge	7 December 1888	375,000
Julia	8 January 1889	150,000
Lautaro	9 January 1889	300,000
San Donato	16 January 1889	200,000
San Sebastian	25 January 1889	160,000
Tamarugal	15 February 1889	650,000
Santa Rita	30 March 1889	100,000
Rosario	15 April 1889	1,250,000
Santa Elena	18 July 1889	110,000
Paccha-Jazpampa	18 November 1889	360,000
	Total:	4,820,000

Source: SAJ, 17 January 1891.

increased in the intervening years.[17] In January 1889, James, Inglis and Company sold the San Donato *oficina* to the San Donato Nitrate Company for £150,000. They had purchased the same establishment one month earlier for £90,000.[18] With profits of at least 63 percent to be made by selling to British companies, it is hardly surprising that even long established firms like Gildemeister and Gibbs disposed of their plants.

The Rosario Nitrate Company, launched in February 1889, consisted of the *oficinas* Rosario, San Juan, and Argentina, purchased from J.D. Gildemeister and Company for £1.2 million.[19] The Gibbs partners in London began discussing the possiblity of selling the *oficinas* La Patria and La Palma in September 1888. In February 1889 Gibbs transferred its two establishments to the Tamarugal Nitrate Company organized by George Inglis, and Henry Read, a former Gibbs partner who now served as the manager of the Bank of London, Mexico, and South America.[20] In preparing for the sale, the house had set a value of £318,000 on the properties. But in 1887 the Gibbs's accounts listed their value at only £200,000. The partners readily admitted that the higher figure represented a price that only a speculator like North would pay. Nevertheless, the Tamarugal Company purchased the *oficinas* from Inglis and Read for £590,000, and they in turn paid Gibbs £550,000. Gibbs thus cleared a profit of approximately 175 percent on the transaction.[21]

The demand for nitrate shares became a veritable mania for many British investors. As the *Financial News* reported:

> The company promoter has only to whisper the magic word "nitrates" and the market rises at him, whether he hails from Antofagasta, Taltal or Tarapacá. Gold can no longer conjure up premiums in Chapel Court [the stock exchange] like nitrate. It is the spell which draws the biggest crowd, and causes the greatest flurry among premium hunters.[22]

The Zout Kom Nitrate Company provides one example of how far investors strained their own credulity in buying nitrate shares. The company was launched with the intention of buying 7,000 acres of nitrate land in the Cape Colony, South Africa. The company's promoters argued that since the property was on the same latitude as Chile, similar atmospheric conditions had created nitrate in the area. A few days after the company's founding a rumor spread that Professor V. V. Bradford, who had reported on the nitrate content of the land, had previously attempted to sell it as an agricultural estate! Furthermore, the good professor's references consisted of a letter of recommendation written by an English nobleman in 1869! Amid the catcalls and jeers of its shareholders, the corporation's promoters quickly moved to liquidate the first and last of the South African nitrate companies.[23] But investors remained undeterred by such fiascos, and the nitrate boom continued.

Following the logic of monopoly capitalism, the nitrate boom quickly spread to the industry's service sectors. Organized in September 1888 with a capital of £400,000, the Tarapacá Waterworks Company purchased and expanded North's water supply enterprise in Iquique.[24] This was followed a few months later by the initiation of the Bank of Tarapacá and London, capitalized at £1 million. The bank planned to extend credit facilities to enterprises operating in the nitrate region.[25] Furthermore, in 1889 North floated the Nitrate Provisions and Supply Company with an authorized capital of £200,000. The object of this firm was to supply merchandise, particularly grain and livestock, to the nitrate *oficinas*.[26] By far the most important of North's service enterprises was the Nitrate Railways Company.

The Montero brothers' railway firm, incorporated in London as the Nitrate Railways Company, had faced hard times after the Chilean occupation of Tarapacá. The railway's high freight charges outraged nitrate producers, and despite the high tariffs, the company compiled a poor profit record. Beginning in 1881 various private interests, including the Campbell Company, petitioned the Chilean government for the right to build competing lines. Finally, in January 1886, President Domingo Santa María, arguing that the Monteros had failed to build a branch line agreed upon in the original concessions from the Peruvian government, issued a decree nullifying their railway monopoly in Tarapacá.[27]

By the mid-1880s, then, the Nitrate Railways Company had become a less than inviting investment prospect. Gibbs considered buying into the company and serving as its consignment agent but rejected both options on the grounds that the firm was unpopular, unprofitable, and might soon lose its monopoly.[28] John Thomas North was far more of a gambler than the Gibbs partners. In the spring of 1886, Bailey Hawkins, an associate of North in the Primitiva Company and a director of the Nitrate Railways Company, reported that the Monteros could not repay a debt of £73,000 to a group of British financiers. In April 1887, North took advantage of the Monteros' plight when he purchased 7,000 of their railway shares at 14 percent of their face value.[29] Although North now offered Gibbs the opportunity to buy into the company, they rejected the offer, citing their earlier decision to forgo such an investment.[30] Despite the risks involved, North once again appeared to have landed on his feet. With the company appealing the nullification decree in the Chilean courts and with North serving as chairman, the firm reported dividends on its ordinary shares of 10 percent in 1888 and 25 percent in 1889.[31] These impressive dividends and the seemingly irresistible attraction of nitrate shares did not simply reflect the industry's profitable operation after the collapse of the First Combination.

As the process of capital centralization accelerated, the individual capitalist was separated from the means of production, opening the way for "a new aristocracy of finance, a new sort of parasites in the shape of promoters, speculators, and merely nominal directors; a whole system of swindling and cheating by means of corporation juggling, stock jobbing and stock speculation."[32] Particularly favored was the practice of "making a market." This refers to various methods of inflating the market price of an issue, usually by prearranging the sale of shares to brokers and stock jobbers, giving the impression of intense public demand for the stock. Once unsuspecting investors began buying at the inflated price, the original promoters could sell their shares at a handsome profit.[33] Such manipulations multiplied as the exchange attracted investors with little knowledge of its workings and capital flowed into foreign enterprises in which even practiced investors had little experience.

Instances of such stock manipulation were numerous in the nitrate boom. The shares of North's Tarapacá Bank were allotted to privileged insiders well before the share lists were open to the public.[34] Investors complained that despite making early applications for stock in North's nitrate companies they were unable to acquire a single share.[35] Although North left no private records of his business dealings, it is also clear that, at least in the case of the Primitiva Company, he secretly sold off his shares while the company's stock was still at a high premium.[36] Even the conservative Gibbs house agreed to take £87,000 in debentures of the Rosario Company and £25,000 in shares of the Tamarugal Company with the expressed intention of working them off upon the public.[37] Gibbs also considered purchasing the shares of other companies to run up the market in preparation for the launching of the Tamarugal Company. The house finally rejected the plan only because nitrate shares already commanded a high premium.[38]

Paying excessive dividends was yet another method of creating a market. North's Primitiva Company paid out spectacular dividends totaling £180,000 between 1888 and 1889, but a study of the firm's own reports demonstrated that the dividends exceeded actual earnings by £50,000.[39] The *Financial News* concluded that:

> Whatever he [North] and his colleagues on the Primitiva Board may do, there is one thing quite beyond their power. They may pay an eighty or

THE EMERGENCE OF MONOPOLY CAPITALISM   119

hundred per cent dividend, but they will not find within a half mile of the Bank any man fool enough to believe they have earned it.[40]

North became a master of such larcenous techniques, but again his conduct was not exceptional. In fact, such operations appear to have been commonplace among nitrate promoters.

Inflation of share prices became an integral part of the profit-making mechanism for both sellers of nitrate properties and promoters of companies. Of the sixteen nitrate-producing companies for which information is available, only four paid for their property entirely in cash; the remainder paid anywhere from 13 to 100 percent of the purchase price with their own stock.[41] And there is little doubt that the spectacular rise in nitrate shares during 1888 and the beginning of 1889 was in large measure an artificial creation of company promoters. The success of these manipulations is further evidence that the nitrate boom was the work of a select circle of financiers.

A further consequence of capital centralization according to Paul Sweezy, is that "the big capitalist who can command a large block of shares in one or more corporations is able to bring under his control an amount of capital several times what he owns."[42] Such was the case in nitrates where promotion and direction of the enterprises remained in a relatively few hands. Between 1888 and 1889 eighty-three individuals composed the boards of directors of the seventeen companies. As Table 14 indicates, interlocking directorships characterized these boards. Furthermore, John Waite, H. B. James, R. R. Lockett, W. J. Lockett, W. MacAndrew, George Petrie, and Robert Harvey were all business partners of North. By 1890 the total authorized capital of British nitrate enterprises registered with the stock exchange reached £10,082,900. Of this amount the companies controlled by North and his associates accounted for £6,786,000.[43] This imposing edifice of monopoly capitalism, however, soon began displaying serious signs of structural weakness.

The nitrate industry had been prone to crises of oversupply long before the appearance of the London companies. However, the European trading houses, and other private *salitreros,* could secure a profit only by producing at a cost lower than the price at which they sold nitrate. In contrast, the London companies were built on inflated stock prices and dividends that had to be maintained if their promoters were successfully to rid themselves of their shares. Maximum production even at a loss could secure revenues for posting impressive if unwarranted dividends. Furthermore, the excessive prices they had paid for their properties left the companies heavily overcapitalized. As the *South American Journal* noted, "The companies, from excessive capitalization must produce to their maximum capacities in order to pay dividends."[44] This necessity of producing at the maximum possible output finally broke the nitrate boom. While world consumption remained high, it simply could not keep pace with production (see Table 11). In April 1889 the *South American Journal* warned that the growing surplus of nitrate in European warehouses was depressing the price of the article.[45] By December 31, nitrate had fallen to £8.7.6 per ton compared with an average price of £9.10 in 1888. The depression continued through 1890 with nitrate averaging £8.14.6 per ton, the lowest annual price since nitrate had become an important commercial product. At these prices, few of the British companies could make a profit.[46]

Table 14
Directors of Nitrate Production Companies Serving on the Board of More Than One Company, 1888–89

Individual	Number of Directorships
Robert Harvey	5
George Bush	5
John T. North	4
R. R. Lockett	4
John Waite	4
George M. Inglis	4
F. H. Evans	3
H. B. James	3
William MacAndrew	3
W. J. Lockett	2
George Petrie	2

Source: BOI, 7, 1889, pp. 704–823; 8, 1890, pp. 758–909.

The race to achieve maximum output also caused an increase in production costs. As one contemporary observer concluded:

> It was necessary to show dividends with the least possible delay. Labour on the spot being all employed, it was obligatory to obtain men from any quarter and at any price... the price of labour rapidly advanced with the increasing demand of the new companies, until 5 dollars; and even as high as 7 dollars per day was paid in their desperate need and rush for results.[47]

These wage scales were four times as high as the ones prevailing only a few years earlier (see Appendix). A decline in productivity further aggravated the problem.

Between 1887 and 1890 output per worker in Tarapacá fell by nearly 10

THE EMERGENCE OF MONOPOLY CAPITALISM   121

percent (see Appendix). Chilean government officials attributed the problem to the declining quality of caliche.[48] However, the London companies probably exacerbated the problem in their desperate attempts to increase production. As a result of their efforts, the labor force in the province had swelled from 7,201 in 1887 to 11,657 in 1890 (see Appendix). Adding thousands of additional workers to essentially the same physical plant no doubt created inefficient redundancies in the production process. Labor added further uncertainty to the condition of the industry when long smoldering grievances of the nitrate workers burst forth in a general strike.

Ironically, in promoting the *gremios de jornaleros* the Chilean government had provided workers in Tarapacá with the first institutionalized worker organizations. A strike by the dockworkers early in July 1890 initiated a work stoppage that quickly spread throughout the nitrate zone. The producers agreed to demands for an end to the *ficha* system and company store monopolies only long enough for government troops to be stationed in the *oficinas* as an effective deterrent to further protests.[49] While of only a few weeks' duration, the strike demonstrated that the increasingly militant workers now possessed sufficient coherence to paralyze the entire industry. Such a prospect hardly served to stabilize the weakening market for nitrate shares in London.

The British companies, having paid exorbitant prices for their properties and starved themselves of working capital, quickly fell prey to the problems of declining prices and rising costs. The directors of the Santa Rita Nitrate Company reported that they could not store the firm's nitrate and wait for a rise in prices because of the company's small working capital.[50] Other firms obtained bank loans to pay for storage of their nitrate, but they soon found the 6 percent interest charge eating into their slender profits.[51] The directors of the Taltal and Santa Luisa companies requested approval of mortgage bonds and debentures to create needed working capital.[52] The Tamarugal Company, considered one of the strongest of the British firms, found its profit margin falling well below anticipated levels, and its chairman demanded that Gibbs return a part of the purchase price and grant the company a loan to refund his stockholders a portion of their share values.[53] During the general strike, Herbert Gibbs sarcastically commented that "we have some hopes Tamarugal may be looted off the face of the earth."[54] The actual solution arrived at by the company was less drastic. The corporation exchanged a portion of its ordinary shares for preference shares and bonds, thus reducing its problems of overcapitalization.[55] The Tamarugal Company actually fared far better than most of its competitors.

The days of spectacular dividends and soaring share prices were buried in an avalanche of unsold nitrate. Of the nine companies issuing annual reports as of June 1890, six reported no dividends; the other three managed to issue dividends of 10 percent or less. By December nitrate shares were being offered at one-quarter or even one-eighth of the prices they commanded in 1889. Shareholders' meetings became the scene of rancorous debates and bitter accusations.[56] While North's cajolery still maintained calm at the meetings over which he presided, even the Nitrate King could not maintain peace in the entire nitrate realm.[57] Investors made public their skepticism about the viability of North's Paccha-Jazpampa Company. On one occasion, when North's broker entered the exchange building, copies of the company's prospectus were piled on the floor and burned.[58] With the nitrate boom disappearing in a puff of smoke, it now became a matter of preventing a decline in profits from becoming a collapse of corporations.

John Thomas North, the archetypical monopoly capitalist, was well aware that inflated dividends and a jovial personality would not sustain the nitrate boom. He had already safeguarded much of his own capital in the Nitrate Railways Company, which enjoyed continued prosperity owing to its monopoly on rail transportation in Tarapacá. Next to sales of overpriced *oficinas,* the railway became the largest revenue producer for North.[59] More important, the company could serve as the basis for the monopoly capitalist's ultimate solution to competitive pressures, a single company to control the entire industry. As early as May 1887, North had proposed using his transportation monopoly to force nitrate producers to sell out to a corporation that he would form.[60] Such a scheme would now be facilitated by the control he exercised over so many of the London nitrate companies. Nevertheless, serious obstacles stood in the way of this ultimate solution. North had to reckon with other powers within the nitrate industry such as Gibbs and Campbell. Furthermore, the key to the entire scheme, the Nitrate Railways Company, was under attack in Chile. As threats to his interests multiplied, North involved himself ever more directly in Chilean domestic affairs, and in February 1889 he departed for Chile to review his investments at first hand and to meet personally with President Balmaceda.

Although the name John Thomas North has become synonymous with the British nitrate boom and its eventual collapse, the mania for nitrate shares was the product of more far-reaching influences than the manipulations of one financier. The London nitrate companies evolved from the phenomenon of capital centralization sweeping British capitalism and continued efforts of the capitalist system to lower the cost of raw materials. As was typical of capital centralization, the nitrate boom was built on a network of stock manipulations that placed control of a vastly expanded block of capital in the hands of a few promoters. The rationale of competitive capitalism to produce at a profit did not rule these corporations, whose principal concern was a sizable dividend achieved by whatever means necessary. These activities led to a frantic drive to raise production levels even by grossly inefficient methods and despite steady price declines. At the same time, these overcapitalized enterprises displayed acute sensitivity to competitive pressures. Stock manipulation, concentrated control, and overcapitalization led to increasing pressure for surcease from competition, ideally in the form of a total monopoly or, failing that, a new combination. Unlike previous crises when the largest producers were also the strongest and could usually rely on their economic position to carry the day, these threatened corporations now turned to political influence to protect their investments.[61]

In the nitrate industry raw materials were indeed supplied to Europe at a lower price, but given the form of the London corporations, that process threatened them with destruction. Trapped in this contradiction, monopoly capitalism—as it would elsewhere—sought a solution in the powers of the state.[62] But in this instance, it directed those efforts primarily at state apparatus on the periphery rather than at the center. This maneuver, which accelerated the involvement of foreign nitrate producers in Chilean political affairs, became part of a larger process that toppled the Chilean presidential system in 1891.

[8]

The State in Transition

THE Chilean Civil War of 1891 that toppled President José Manuel Balmaceda (1886–91) from power marked a decisive turning point in the nation's history.[1] The Civil War, which pitted the legislature against the executive, ended over a half century of rule by executive regimes. Its destruction of the presidential system seemed to afflict the body politic with an incurable malaise. In the ensuing era of congressional dominance (1891–1924), the government often squandered its revenues, social conflict reached crisis proportions, and the state was wracked by petty political bickering and ineffectual policymaking. Despite its apparently dire consequences, the Cival War represented an adjustment in state structures essential to the continued development of Chile's new relationship to the capitalist center. After the capture of the nitrate regions in 1879, the state served as the oligarchy's key link to the nitrate industry and was the most vital force in the domestic economy. This new role of the government rapidly increased political tensions. The elite splintered into innumerable factions vying for a share of state resources and seeking to protect existing relationships to the nitrate industry. The penetration of monopoly capitalism exacerbated these conflicts as Chilean politicians became the partisans of competing European nitrate producers. In the Civil War of 1891 these forces converged on and destroyed the Chilean presidential system, the one state institution that threatened to disrupt this process.

The Constitution of 1833 conferred sweeping powers upon the executive. Eligible for two five-year terms, the president had wide-ranging appointive authority that assured him control over the electoral process. Chief executives systematically rigged elections to ensure a favorable congressional majority, secure their own reelection, and impose a handpicked successor on the country. If despite electoral intervention congress should prove uncooperative, the president could fall back upon his power of absolute veto and the right to close congress. More serious challenges from outside the legislature could be met with the declaration of a state of siege and through the president's control of the armed forces.[2] This system of constitutional despotism suited the conditions that prevailed in Chilean society in the first half of the nineteenth century.

Until at least mid-century, Chile remained a relatively isolated agrarian society dominated by a close-knit landowning elite. The functions of the state were essentially confined to defense against external aggression and preservation of internal order. The latter proved an unlikely source of tension in a society where open conflict along class lines was virtually nonexistent. Moreover, the power of the state stopped at the gates to the great estate where the hacendado neither needed nor welcomed state interference. In a society with the vast majority of its population comprised of underemployed rural inhabitants lacking villages as centers of resistance, and capped by a homogeneous ruling class, near dictatorial presidential rule became an acceptable means of resolving disputes that tended to erupt along familial or regional cleavages. But by the 1850s, signs of change were already apparent.

Despite the predominant role assigned to the presidency, the framers of the 1833 Constitution had not consigned congress to total subservience. The legislature's most important power was budget approval. As early as 1841 congress had forced compromise on the executive by refusing to vote on tax collections and the budget for the coming year.[3] Efforts to advance beyond this original constitutional check upon the presidency emerged as state services multiplied.

Growth of Chile's international trade poured steadily increasing revenues into government coffers. Even a state committed to laissez-faire policies must undertake projects and reforms to create optimum conditions for growth of a market economy.[4] These changes became apparent during the presidency of Manuel Montt (1851–61) in an array of public works projects and steady growth of the bureaucracy. Moreover, an 1854 law gave executive appointees a decisive role in local government, further extending presidential influence at the municipal level. Rationalization of state functions brought the Montt administration into conflict with the Catholic church as the executive asserted the supremacy of civil over religious authority. The dispute unleashed a decades-long struggle to curtail church influence in Chilean society. The religious issue was symptomatic of a more general trend in the political process.

As the power and influence of the state penetrated ever more deeply into the fabric of Chilean society, the elite sought to curtail the executive's domination of governmental apparatus.[5] The church dispute itself limited presidential authority. Owing to the religious issue, the ruling conservative coalition split into pro- and anticlerical factions, the Conservative and National parties. The opposition Liberals had also fractured with more extreme anticlerical members forming the Radical Party. Thus, Montt's successors found it necessary to rule through a series of legislative coalitions.[6] Conscious attempts to restrict executive power accelerated in the 1870s. Legislation passed in 1871 limited the president to a single five-year term, and the 1874 suffrage law made it more difficult, though by no means impossible, for the president to control elections. Although couched in the phraseology of political liberalism, the debate on executive authority did not represent the struggle of an emergent middle class against a landed aristocracy. As Arturo Valenzuela has noted concerning the 1874 suffrage law:

> The law was not the creation of Liberals seeking to curb the excesses of an authoritarian state in the hands of Conservatives but of a coalition seeking to curb the power of the presidency. As in several European countries, the coalition included Conservatives who believed that a freer suffrage system would be to their advantage because of the control which local notables had over the political behavior of elements of the lower class.[7]

Even as merchants and miners joined the elite, it remained united in its basic socioeconomic interests, and it was within the oligarchy itself that the debate raged. If advocates of a limited executive had a democratic ideal, it was to circumscribe the control of an ever more powerful executive over the affairs of the elite. One of the most ardent of those advocates was José Manuel Balmaceda.

Son of a well-to-do landowner, Balmaceda from the outset of his political career enthusiastically supported limits on executive electoral interference and on the influence of the church. After his election to the Chamber of Deputies in 1870, he continued to champion these reforms and press for legislation to enact them. As minister of the interior under President Domingo Santa María (1881–86), Balmaceda oversaw implementation of laws laicizing cemeteries and establishing civil registries for births and marriages. However, his attitude toward electoral interference underwent a marked change. From his influential cabinet position he engaged in the manipulations necessary to assure his election as Santa María's successor. His volte-face on electoral interference was hardly exceptional. The transformation from liberal reformer to autocratic executive was quite common among Chilean politicians of the period.[8] The possibility that Balmaceda could uphold the tradition of strong executive rule seemed strengthened by the comfortable congressional majority he enjoyed at the outset of his presidency. Subsequently, executive involvement in the balloting of 1888 contributed to the election of a legislature in which 109 of 123 deputies and 27 of 28 senators supported his administration. While intermarriage among the elite had always ensured that a large number of congressmen were related by blood, the 1888 congress displayed the highest concentration of familial linkages of any legislature since 1834. Thus, one of the basic adhesives holding the Chilean political process together appeared to be stronger than ever.[9] Yet by 1890 Balmaceda's congressional majority had disintegrated, and he rejected the resignation of his cabinet, despite its censure by the legislature. Finally, when congress refused to approve the tax collections for 1891, Balmaceda decreed that the 1890 taxes would remain in effect. In response, congress rose in revolt, plunging Chile into civil war.

In searching for a catalyst in the crisis of 1891, historians have identified exceptional qualities in Balmaceda, either as an idealist committed to economic development that threatened the interests of the elite, or a short-sighted egotist unwilling to persist in the political gamesmanship necessary to maintain a legislative majority.[10] But neither image provides a completely adequate explanation for his overthrow. His development programs, designed to expand Chile's economic infrastructure, differed in scope but not in intent from those of his predecessors. Extension of transportation and communication facilities serviced existing agricultural, mining, and commercial interests. The proliferation of new schools did little to alter an educational system designed to prepare the children of the well-to-do for professional careers. Balmaceda himself reportedly said that "equal distribution of public jobs and the suppression of direct and sub-alternate taxation... has constituted the foundation of the economic policy to which I have been conforming in a gradual and constant fashion."[11] In fact, during his administration the share of government revenues contributed by direct taxes on wealth declined by more than 50 percent.[12] Such policies could hardly be construed as representing a fundamental challenge to the ruling elite.

Balmaceda's self-image and sense of historic mission were nothing short of heroic. Yet with the exception of Aníbal Pinto, Chile's presidents were not renowned for self-effacement. Soon after the installation of Balmaceda's administration, it won praise for its nonintervention in municipal elections held in November 1886. In congress, Balmaceda hoped to rule through a broad coalition including Government Liberals, Nationals, and those elements of the Liberal majority who had opposed his candidacy, the Dissidents. Moreover, his term of office signaled a period of reconciliation between church and state after the rancorous events of Santa María's presidency.[13] As for the intervened congressional elections of 1888, they were notable for a reduction in the violence that had marred previous campaigns. When opposition mounted to Balmaceda's choice of Enrique Sanfuentes as his successor, the president took steps to remove Sanfuentes as a potential candidate.[14] Yet these conciliatory gestures failed to halt the drift toward civil war. Thus, the relatively orderly process by which the elite sought to limit the executive collapsed under a president schooled in the mechanics of the Chilean political system and intent on rule by coalition. The destruction of Balmaceda's regime did not result primarily from radically new economic policies or personal idiosyncrasies of the president. It was rather the product of the new role thrust upon the state during the Nitrate Age.

The most obvious impact of the nitrate industry upon government was the enormous increase in its resources. With the nitrate tax providing more than 50 percent of the total by 1890, government revenues in the decade 1881–90 exceeded £87,600,000, an increase of 74 percent over the period 1871–79.[15] The resulting growth in expenditures intensified the state's control over society. And while congress possessed the power of budgetary approval, the executive managed the dispersal of those funds. During the Santa María administration, sharp criticism was leveled at disbursements for state railways on the ground that they enhanced the political power of the president.[16] Critics aimed similar charges at the expansion of public employment. In fact, by 1884 twenty-four members of the Chamber of Deputies held positions in the bureaucracy to which they had been appointed by the president. A legislative effort in 1886 to ban congressmen from receiving government contracts offers further evidence of the increased power wielded by the executive.[17] The Chilean historian Francisco Encina described the Government Liberals supporting the president in congress as a collection of public employees, their dependents, and government contractors. According to Encina, their loyalty to the administration "was a passive obedience imposed by employment, contract or dependence on the authorities."[18] Increased state revenues tended to counteract the trend toward curtailing the authority of the presidency. This involved more than an exacerbation of the conflict over the power of the executive. If increased state expenditures aided the president in electoral manipulation and control of congress, it also opened vast new opportunities for the elite to extract wealth from the state.

Prior to 1879 the government oversaw an economy based on domestically owned agricultural and mineral resources and the international commerce linking Chile to Europe. While the state's resources rose steadily with the expansion of international trade, wealth produced in the private sector far outdistanced those resources. State revenues in the decade preceding the War of the Pacific equaled less than 57 percent of the value of total exports. The balance of the national economy was thus heavily weighted in favor of the private sector. With the incorporation of the nitrate industry, control of the nation's most important resource passed into foreign hands. In the decade 1881–90, nitrates accounted for just over 50 percent of Chile's total exports. Of 422 million pesos in revenues produced by the nitrate industry, approximately one third was retained by segments of the domestic economy in the form of payments to labor, consumption of national products, and other factors. But the state assumed the single most important role in channeling nitrate wealth into the economy, retaining an additional third of those revenues through its nitrate and iodine taxes.[19] The importance of this development was reflected in the reversal of the ratio between the public and private areas of the economy. While the private sector now enjoyed its own linkages to the nitrate industry, this was coupled with the faltering performance of mining and agricultural exports. As a result, state revenues equaled 111 percent of private domestic revenues (nonnitrate exports plus returns to the private domestic economy from the nitrate zone) in the period 1881–90 (see Table 15). The resources controlled by the elite had not totally disintegrated, and agriculture at least had an expanded domestic market to compensate for the drop in exports. But nitrates had assumed primacy in the economy through government nitrate taxes and market linkages to the domestic economy. And, with the state emerging as the single most important link to the new wealth, the political arena became a focal point for economic competition.

Public revenues flowed into the domestic economy in a variety of ways. Nitrate revenues deposited in banks were loaned out to private interests.[20] Public works projects brought profits to contractors as well as employment and long-term benefits such as improved transportation to the regions in which they were initiated. The state bureaucracy, which expanded and

Table 15
State Revenues and Private Domestic Revenues (Current Pesos), 1881–90

Year	State Revenues	Private Domestic Revenues[a]	% State/Private
1881	40,133,262	43,673,559	92%
1882	42,783,158	51,060,354	84
1883	47,750,545	57,764,709	83
1884	38,892,990	42,423,517	92
1885	39,438,668	38,694,105	102
1886	60,543,250	40,229,798	150
1887	68,308,136	43,810,906	156
1888	52,096,734	55,061,340	95
1889	62,209,269	46,197,501	135
1890	59,218,710	48,842,078	121
		Average:	111%

Sources: *Hacienda pública*, pp. 26–29; Chile, Ministerio de Hacienda, *Memoria de 1890–1891* (Santiago, 1891), p. 25.

[a]These figures were arrived at by combining total non-nitrate exports and estimated returns to the domestic economy from direct linkages to the nitrate industry. Estimated returns are based on Markos Mamalakis's calculation that approximately 33 percent of nitrate revenues were retained by the domestic economy. This does not include the 33 percent absorbed by state taxation. Since the estimate is an average for the entire Nitrate Age and assumes substantial Chilean ownership of the industry, actual returns to the national economy for the period 1881–90 were no doubt significantly less than one-third of nitrate revenues; see Markos J. Mamalakis, "The Role of Government in the Resource Transfer and Resource Allocation Processes: The Chilean Nitrate Sector, 1880–1930," in Gustav Ranis, ed., *Government and Economic Development* (New Haven, Conn., 1971), pp. 181–210.

offered increased remuneration to officeholders, symptomized the new role of the public sector in the Nitrate Age.

Traditionally, sons of well-to-do Chileans had accepted government positions for the prestige they conveyed. As late as 1878 the Chilean diplomat, Alberto Blest Gana, noted that the scions of wealthy families would be willing to accept unpaid positions in the bureaucracy in order "to enrich themselves with honor."[21] But during the Nitrate Age bureaucratic posts took on a new significance as a source of income. In his hypercritical assessment of Domingo Santa María's presidency, Carlos Walker Martínez described the growth of public employment in the following terms:

> The craze for public office developed in so vast a scale around the government serving the men of power and with a veritable frenzy in regard to the budgets, that the state can well be compared to those large trees ensnared by numerous vines and parasitic plants which have their vitality sapped by the detestable nodes of their snares.[22]

Resurgent government finances also enabled the state to assist specific areas of the economy. Increased expenditures on rail transport and the expanded credit facilities offered by the Caja de Crédito Hipotecario aided agriculturists in the transition from export to domestic markets. The most needy recipient was the mining region where copper production fell by 33 percent between 1881 and 1890.[23] The inadequacy of the region's privately owned rail system posed a serious threat to its economic survival. In 1887, with the bankrupt Chañaral railroad about to be dismantled, the state responded to the pleas of local mineowners and purchased the line for 350,504 pesos. In the following two years the government began construction of two new lines in the mining region, and plans were under consideration to expend 6 million pesos acquiring additional mining railways.[24]

Economic interest groups had traditionally sought to protect their positions through political action such as debates on tariffs, but political activity now took on a new meaning as elite factions vied for a share of state wealth. The process of protecting private sector investments from state extractive functions became secondary to efforts by private interests to extract state resources.

Before the end of Domingo Santa María's term of office, questions were raised concerning expenditures such as the 1.7 million pesos dispersed for a drydock for which a site had not even been selected.[25] Santa María's most caustic critic described this process of extracting state resources in the following terms:

> Extravagance was converted into a system of government, and favoritism spilled the wealth of the nation into the pockets of friends. The railroads, wharves, public buildings, official pressworks, customhouses, guano, nitrate, the intendencies, governorships, the police, all were no more than an unbridled embezzlement on which they threw themselves like vultures on their prey, the palace adulators, the intimate friends and relatives of the President of the Republic, and friends and relations of the friends, provided they yielded to the approved official servitude. And so after those two hundred million pesos were squandered one asks with horror, What has been done? What are and where are the works of manifest usefulness which they have produced? Little of use has remained, and what has remained, is not worth half of what it costs, the other half is the surplus that belonged to the patron saints of the great chaplaincy into which the administration of 1881 to 1886 converted the Republic of Chile.[26]

Nitrate wealth had created a fundamental contradiction within Chilean political institutions. While opening a variety of opportunities for enrichment from state coffers, it reinforced the executive's control of the political process and thereby state wealth as well. During the Santa María administration the potential existed, and was partly realized, of creating a presidential constituency from the resources of the state. As opportunities for wealth increased, so did the possibility of limiting its distribution to the political adherents of the executive. These conflicting forces came to a head during Balmaceda's presidency.

Balmaceda's first two cabinets represented alliances between Government Liberals and the National Party. It soon became apparent, however, that the dual effects of nitrate revenues, increased executive power, and politicization of economic competition would doom efforts to maintain a stable coalition. From the outset, the Independent Liberals, or Dissidents, feared the consequences of their exclusion from the cabinet. Lacking strong local constituencies, the Dissidents believed reelection to congress would be impossible without presidential support. The tranquility of the 1888 elections is moot testimony to the accuracy of their concern. For the reduction in violence did not reflect political harmony but the enhancement of executive electoral control stemming from the increased funds and number of appointive positions at the president's disposal. In the newly elected congress, fifty members held public positions by executive appointment, nearly twice as many as in 1885.[27] The Dissidents were also haunted by the realization that lucrative public works, government contracts, and positions in the bureaucracy would accrue to the Government Liberals and Nationals who controlled the executive branch. Thus, for the Dissidents, admission to the cabinet and expulsion of the Nationals became a life-or-death struggle.[28] The competition for state revenues became a source of bitter contention even within the government coalition.

Despite a presidential majority in the legislature, congress in 1887 rejected Balmaceda's initial proposal to buy the Chañaral railway. The opposition resulted from the fact that the president's minister of finance, Agustín Edwards, the principal shareholder in the railroad, would earn an exorbitant profit at the privately agreed on price. Only when the president consented to a lower figure did congress pass the measure.[29] Balmaceda's utilization of state revenues also set off other acrimonious disputes.

Since 1873 the Banco Nacional had been the exclusive banker for the government. In 1888, the executive reached a new agreement with the bank allowing the state to deposit its funds in a number of different financial institutions. Balmaceda argued that this move would benefit the entire economy by keeping interest rates down. Initially, Deputy Enrique MacIver assailed the plan on the ground that it gave the executive excessive influence over the national credit system. Once in operation, the project's critics became more specific. They argued that fiscal deposits were being made in only three Santiago banks, favoring them over other institutions.

They assailed deposits made in a fourth bank, the Popular Hipotecario, because the bank's manager was a political ally of the president. Opponents also attacked the policy on the grounds that the banks dispersed the funds to a few favored customers and that tying up public revenues in this manner delayed important projects such as the purchase of mining railroads. Finally, in July 1889, with the state's deposits nearing 18 million pesos, the minister of finance agreed to begin withdrawing the funds from the banks.[30] Public employment became yet another area of controversy.

Not only did the size and salaries of the bureaucracy increase under Balmaceda, but critics accused the president of creating new positions without congressional approval. Since appointments to such positions were entirely the prerogative of the president, some of Chile's leading families sought Balmaceda's assistance in obtaining appointments or promotions for their relatives. Once again, competition among the elite brought forth charges that the president created and filled such posts to allow a few favored individuals to enrich themselves.[31] Moreover, elite concern with the growing power of the executive focused not only on the dispersal of public revenues but on state influence over the nitrate industry as well.

Although the Chilean government had pursued a strict laissez-faire policy toward the nitrate industry since the issuance of the decree of June 11, 1881, the state inevitably assumed an important regulatory function over the nation's most valuable natural resource. The Santa María administration's granting of port facilities to the Campbell Company for the export of nitrates and its nullification of the nitrate railway monopoly in Tarapacá clearly indicated the ability of the state to influence the industry's development.[32] During the 1880s, the government's holdings of unused nitrate lands took on greater significance as the older *oficinas* in Tarapacá exhausted their nitrate deposits and as the number of private properties available for sale dwindled. Balmaceda strengthened control of public nitrate reserves with creation of the Delegación Fiscal de Salitreras y Guaneras to oversee the nitrate properties. In attempting to limit the extent of foreign control in the nitrate industry, President Balmaceda exploited state power to the fullest.

In April 1887 the legislature authorized the executive to contract a foreign loan of £1,113,781 to purchase unceled nitrate certificates held by Italian and German investors. This action added additional nitrate properties to the tracts already held by the government. After congress approved the legislation, rumors spread that the executive planned to use state properties to organize a national nitrate company.[33] In January 1888 Balmaceda informed Brice Miller, one of the Gibbs partners, that he intended to reduce foreign control by forming just such a corporation. The state would sell its nitrate holdings to the firm, which would also buy privately owned nitrate properties such as Gibbs's Alianza grounds. It was believed that private interests would be willing to sell at low prices because of the burden of the railway monopoly. Once the company was formed, the state would enforce the 1886 nullification decree against the railroad.[34] Balmaceda thus focused the considerable powers of his office on the task of increasing Chilean investment in the nitrate industry.

The president's plans quickly aroused opposition from elite factions intent upon protecting or contending for a share of the nation's nitrate wealth. Gonzalo Bulnes, the former intendant of Tarapacá, argued that foreign development of the nitrate industry had aided Chilean agriculture and Chilean banks. He went on to state that:

> I don't wish to concern myself now with examining the means which, according to public rumor, the government is considering in order to nationalize the wealth of Tarapacá. If what is being said is true, they would demolish the Chilean nitrate industry established in Taltal and in Antofagasta; they would place in serious peril the Chilean capital of the banks distributed among the *salitreros* and would only manage to produce a disastrous struggle.[35]

Any threat to the banks would be particularly unpopular, since, as Bulnes noted, thousands of their stockholders profited from the banks' relationship with the nitrate industry. Additional opposition stemmed from the intention of Balmaceda's minister of finance, Agustín Edwards, to play a major role in the national company. Having just defeated the initial Chañaral railway proposal, the legislature was not about to approve any other scheme that might add to the Edwards fortune.[36] Opposition to Balmaceda's plan resulted from the fear that it would disrupt the beneficial relationship between the industry and the domestic economy and from the belief that it would serve the personal interests of Agustín Edwards, the darling of the National Party.

The president's next plan proved to be far less ambitious. In June 1888, the executive requested from congress the right to auction off the most recently acquired state nitrate lands during a three-year period. Although the final version of the proposal provided for easy payment terms to encourage Chileans to bid for the grounds, the project obviously did not offer the same prospect for nationalizing the industry as did the national company plan. It did possess the advantage of not favoring any specific interest group, and the Senate approved the measure in August.[37]

By the beginning of 1889, both the president and private interests had good reason to reconsider the auction plan. Between August 1888 and February 1889 nine new nitrate companies appeared on the London stock exchange. As Balmaceda soon pointed out, his concern was not simply foreign predominance but a monopoly in the hands of investors of a single nation.[38] The president's sentiments now enjoyed a degree of popular support.

On several occasions during the 1880s, the Gibbs partners noted the existence of antiforeign feelings in Chile and, in particular, opposition to foreign monopolies.[39] Such observations might appear incongruous with the encouragement given to foreign investment in the nitrate industry. Yet the apparent contradiction represented a very rational attitude of the elite. Foreign investment was sought in those areas such as nitrates whose development lay beyond the capabilities of the domestic socioeconomic system. Conversely, opposition readily emerged when European capitalism threatened to overrun those endeavors the elite considered within its purview. During Balmaceda's presidency, for example, congress began to formulate legislation to restrict the activities of foreign joint-stock companies operating in Chile. Of particular concern were the inroads made by European insurance firms that had gained control of nearly 80 percent of the insurance market by 1883.[40] A similar situation now appeared to be developing in the nitrate industry.

In December 1888 the Bank of Tarapacá and London came into existence for the purpose of supplying credit to *oficinas* in Tarapacá. It was soon followed by the Nitrate Provisions and Supply Company, which would provide merchandise, including imports of Argentine agricultural products, to the nitrate region. Both of these enterprises represented obvious challenges to the remunerative relationships that Chile's financial, commercial, and agricultural interests enjoyed with the nitrate industry.[41] Calls for a more vigorous nationalization policy now echoed through the press.

The newspaper *La Libertad Electoral* warned that the British monopoly threatened Chile's very sovereignty and castigated Balmaceda for not following through on his original nationalization plan. A recognized expert on nitrates reiterated fears of a total British monopoly and urged the state to establish its own monopoly including a national company with a single enormous *oficina*. The newspaper *La Epoca* called for creation of a national company under state auspices. An apparently well informed source, writing in *El Ferrocarril,* criticized Balmaceda for allowing the British monopoly to reach such dangerous proportions during his administration. The author warned of a foreign production monopoly and repeatedly expressed concern that the new British service companies posed a threat to the linkages between Chilean banking and agricultural interests and the industry. The writer concluded by urging the state to use its nitrate lands and financial resources to create Chilean-owned nitrate companies with nontransferable stock.[42]

With support growing for a more vigorous policy of Chileanization, Balmaceda allowed the auction plan to languish; and in his annual message to congress on June 1, 1889, he proposed the sale of nitrate lands directly to Chileans. Avoiding the drawback of his original national company scheme, which might have favored a small clique, the president now suggested that a number of national companies be formed with nontransferable shares.[43] The new proposal raised a problem that had not existed in the original project. While the nitrate railway monopoly would have facilitated the launching of a single large national company, its high freight charges would clearly discourage the organization of a number of small firms based solely on public lands. In addition, Balmaceda's programs for internal development depended on the maintenance of high government revenues. Again a major obstacle was the Nitrate Railways Company, whose high freight charges, it was believed, restricted exports and consequently the amount of export duty collected.[44] But Balmaceda would soon discover that while the oligarchy did not oppose limitations on foreign investment per se, such restrictions inevitably affected members of the elite affiliated with foreign interests.

As minister of the interior under Santa María, Balmaceda had been an outspoken opponent of the rail monopoly and had urged Santa María to sign the nullification decree in 1886.[45] With the future of his own programs now at stake, Balmaceda became all the more intent on breaking the monopoly. In 1887 the Supreme Court passed favorably on the Nitrate Railways Company's request for judicial review of the nullification decree. As long as the Nitrate Railways Company had served as an unwitting ally in his nitrate company scheme, Balmaceda had taken no further action on the matter. Now that the company stood in the way of his nationalization plan, he referred the case to the Council of State, an advisory body whose members were appointed by the president. The Council in turn denied the right of the court to review the case.[46] By December 1889, Balmaceda, satisfied that the railway monopoly had been legally nullified, made clear his intention to grant a competing railway concession to the Campbell Company.[47]

At this point, the Railways Company appealed to the British Foreign Office. In February 1890 the British minister in Chile, J. G. Kennedy, lodged a formal protest against attempts by the executive to abolish the railway monopoly, but Balmaceda rejected the protest. Subsequently the Foreign Office assumed a more cautious stance when the Gibbs house, which favored breaking the monopoly, protested against its intervention.[48] By April, the Railways Company had initiated a new court action to prevent the executive from granting a concession to Campbell. In addition, the company received assurances from the Balmaceda administration that it would be permitted to build a rail extension to John Thomas North's nitrate grounds at Lagunas. But even before the diplomatic protest and the appeal to the Chilean judiciary were formulated, the company had initiated a political effort to preserve its monopoly.

In December 1889 the Railways Company sent a memorandum to the Chilean Senate challenging the authority of the Council of State to prevent judicial review of its case. It also suggested that congress could stymie the executive's policies on the nitrate railways by presenting accusations against the president's ministers. Thus, the Railways Company deliberately sought to involve itself directly in Chile's domestic political affairs.[49] That effort signaled a new stage in the relationship between European nitrate interests and the Chilean government.

The nitrate industry drastically altered the relationship between the Chilean state and foreign economic interests. Prior to 1879, British interests usually confined themselves to commercial activity. Operating under a government that created a favorable environment for foreign trade and limited its involvement to the imposition of import and export duties, British merchant houses had little reason to involve themselves in domestic politics. After the occupation of Tarapacá, however, British interests included substantial direct investments in Chile's most important natural resource. And the Chilean government clearly possessed the ability to influence the industry's course of development. Granting rights to port facilities and concessions for rail lines were only two examples of this power. Efforts to influence these policies by appeals to the Foreign Office proved ineffective for several reasons. Chile had escaped the treacherous financial and wartime conditions that had left her so vulnerable to foreign pressure after the seizure of the nitrate regions. Moreover, British nitrate investment consisted of a series of competing interests, unlikely to act in unison in such appeals. As Gibbs's involvement in the railway question suggests, competitive struggles effectively diffused the impact of Foreign Office pressure. But a viable alternative for influencing government policy existed in the use of domestic lobbyists. That process became apparent as the controversy over the Nitrate Railways Company reached its peak.

Foreign capitalists had long made a practice of hiring Chilean lawyers.

Since the law was a common profession among Chilean politicians, many of these attorneys were also prominent political figures.[50] But the deliberations of the Gibbs partners in 1890 on the selection of an attorney indicate a knowledge of the law was not the primary concern of foreign employers. At that time, Gibbs considered two candidates, Senator Eulogio Altamirano, and another member of the legal profession, Horacio Zañartu. The new lawyer was to secure a railway concession for Gibbs's Alianza grounds from the Chilean executive. In discussing the options, one of the partners, Henry Giles Daubney, stated:

> If from what we learn in Santiago, we can trust Mr. Zañartu we shall adopt your suggestion and tell Mr. Altamirano, if necessary, that we are going to work our tender through others; as he, in his capacity of Consejero de Estado will thus doubtless be able to assist us more than he could were he directly "implicado." We can also point out to him that if subsequently we decide to also try to get a concession through Congress we shall submit the whole of such work to him, and as affairs are at present, Mr. Altamirano is doubtless very influential in the Senate.[51]

When Gibbs learned that their tender would not be subject to review by the Council of State, they chose the more politically influential Altamirano as their attorney.[52] As Daubney's remarks indicate, political influence was a primary concern of foreign investors in selecting a Chilean attorney. His statement also implies that foreign employers, wary of exposing their Chilean agents to attacks as tools of foreign interests, preferred, if possible, to employ them in an informal capacity. Gibbs and other private partnerships resorted to such tactics as a response to state control over their investments, and as a competitive reaction to wide-ranging influence peddling by the archmonopoly capitalist, John Thomas North.

Balmaceda's congressional opponents on the railway question included deputies Julio Zegers, chief counsel for the Nitrate Railways Company, and Luis Martiniano Rodríguez, another of John Thomas North's attorneys. In fact, the number of prominent politicians working as legal advisors to North totaled more than a dozen.[53] This impressive collection of public figures is understandable in view of North's position. He had tied up much of his capital in the nitrate railway, the key to a potential production monopoly for shoring up the fragile nitrate companies he had created in London. No doubt political and other considerations influenced the actions of the Chileans. Carlos Walker Martínez, one of North's attorneys, had long been a bitter political opponent of Balmaceda. But it is hard to imagine that they could draw a clear distinction between their interests as the representatives of foreign investors and their role as Chilean legislators.[54] Julio Zegers, for example, was said to have been a friend of Balmaceda who turned against him when the nullification decree threatened his source of income.[55] And as Daubney's comments indicate, foreign investors hired these men specifically for the purpose of exercising their political influence on behalf of their employers. Furthermore, the supporters of the president had similar commitments.

Despite strong congressional opposition, Balmaceda granted the Campbell Company its railway concession. A few days later the British minister pointed out that Balmaceda's favorable action resulted from the fact that:

> Some five or six members of the Chamber of Deputies so far supporters of the President are interested in the success of the above Company [Campbell]: had the Government decided in favour of the Nitrate Railways Company, the above Deputies would have joined the Liberal Alliance.[56]

The concession was the first product of an elaborate series of maneuvers undertaken by the Campbell Company to secure influence in the domestic political process.

In 1889 the Peruvian Guillermo Billinghurst, in one of the best-known works on the nitrate industry, *Los capitales salitreros de Tarapacá,* castigated both North and Gibbs. He also went to absurd lengths to picture the Campbell Company as the work of native, that is, Peruvian and Bolivian capital. Billinghurst was a business associate of the Campbell interests, and his book set the stage for an interesting transformation of the company.[57]

Lauro Barros, a member of the Chamber of Deputies, had been instrumental in securing the railway concession the company received on March 19, 1890. Three days later Campbell transferred the concession to Barros and another Chilean, Pedro Wessel. Subsequently the company was reorganized as the Compañía de Salitres y Ferrocarril de Agua Santa. Wessel and Barros transferred the concession to the Agua Santa Company and became two of its directors, while the Campbell interests retained control of 67 percent of its stock. The transaction was approved by the Balmaceda administration in which Barros now served as minister of finance.[58] The Campbell Company thus secured a vital competitive advantage through the influence of a Chilean politician. At the same time, the elaborate gyrations it undertook prevented Barros from being directly "implicado," while its cosmetic reorganization gave it the appearance of a Chilean corporation.

These events typify the power parameters of the Chilean political system. North, his interests represented by members of the congressional opposition, found his railway monopoly effectively broken, while the Campbell Company, whose agents supported the president, had won its long-sought-after concession. Under pressure from the nitrate crisis, North abandoned his original plan for a total nitrate monopoly, substituting a more easily implemented scheme for a trust to control marketing and production. But Balmaceda also frustrated this plan by dangling a railway concession before Gibbs. In hopes of obtaining a railway for their Alianza grounds, Gibbs refused to join the trust, denying it the financial backing necessary for its formation.[59] The political implications of the Agua Santa case ran even deeper.

In October 1890 Billinghurst wrote to Lauro Barros describing political conditions in Tarapacá in light of the upcoming presidential election. David Mac-Iver, whose brother was a lawyer for the Nitrate Railways Company, led the opposition to the Balmaceda adminstration. His ardent supporter was John Dawson, manager of North's Tarapacá bank. Backing the president were the Agua Santa Company, the Banco Mobilario, and Folsch and Martin. The connection between the three is not difficult to explain. Lauro Barros was a business associate of Melchor Concha y Toro and Francisco Subercaseaux. It was their Banco Mobilario that had opened a 1-million-peso credit account for the Agua Santa Company at the time of its incorporation. Furthermore, the bank served as the principal financial backer for Folsch and Martin.[60]

Billinghurst's comments illustrate the complex network of associations developing between foreign and domestic interests. In seeking to break the railway monopoly, Balmaceda found himself embroiled in the competitive struggle between these groups and ended by allying himself with one faction. That decision had far-reaching political implications, winning the temporary allegiance of Chilean politicians interested in Agua Santa but antagonizing those affiliated with North.

The political consequences of his policies frustrated and angered Balmaceda. His grandiose plans to develop Chile's economic infrastructure and increase Chilean participation in the nitrate industry had set off endless disputes among elite groups anxious to share in state revenues, protect linkages to the nitrate zone, and advance the interests of competing foreign producers. As the principal arena for those disputes, the Chilean congress became the main target of the President's enmity. In June 1890 Balmaceda denounced parliamentary forms of government "as one of those anarchic systems in which personal circles continue to divide, either to support or overthrow ministries, or to serve interests other than those of the common weal"; and he rejected "the bastard parliamentary government of the Republic."[61]

Balmaceda's difficulties were readily apparent in the history of his attempts to rule through a broad-based coalition. Beginning with the alliance of Nationals and Government Liberals, the president subsequently included the Dissident Liberals only to lose the support of the Nationals. There followed ministries composed of only Government Liberals; cabinets that included at one time or another Radicals, Dissidents, and Nationals; and even an apolitical ministry. As elite factions sought to tap state wealth, protect existing links to the nitrate industry and serve foreign nitrate producers, the fabric of Chilean political life became a crazy quilt of conflicting interests. Agustín Edwards, the National Party's leading personality, had attempted to use his position in the executive to expand his interests in the nitrate industry. The Liberals Lauro Barros and Melchor Concha y Toro were affiliated with the Campbell Company; the political affiliations of North's partisans spanned the political spectrum, including Julio Zegers of the Liberals, the Conservative Carlos Walker Martínez, and Enrique Mac-Iver of the Radicals. The process atomized a political system once divided primarily by personal allegiances and questions of presidential power. This acute factionalization doomed attempts by the executive to create a stable ruling coalition. In its six years, the Balmaceda administration went through thirteen complete cabinet changes, more than twice as many as any previous administration. Moreover, by the end of 1889 the president had lost his ruling majority in congress.[62] If the elite was increasingly factionalized by competition for state economic powers, it was at least in agreement that the executive must no longer be allowed to influence that struggle.

In January 1890 the congressional opposition defined its position in a manifesto calling for an end to executive interference in elections, municipal reform, and acceptance of parliamentary government. The second point related to the first, since it implied local versus executive control of the electoral process.[63] The manifesto clearly drew the battle line over presidential control of the political system. Yet the conflict did not represent a pure clash of liberalism versus authoritarianism. To the Conservatives, the only group that had consistently opposed Balmaceda and supported electoral liberty, the battle between the president and his liberal opponents was not "a struggle between liberty and [presidential] intervention but between the ambitions of the distinct circles and caudillos to make the government's influence turn in their favor."[64] The urgency of the conflict derived from the critical link that the state represented between the nitrate industry and the domestic economy. For the state had assumed a vastly expanded economic role that contemporary observers fully appreciated.

In April 1889, Ambrosio Montt, fiscal of the Supreme Court of Chile, delivered a legal opinion concerning a claim to nitrate grounds in Taltal. In that opinion he stated that:

> The government is assuming today, ... a large part of the industrial and economic activity of the republic; and as the administrator of the national patrimony, now ample and rich, as the promoter of fiscal works of public interest, it finds itself carrying foward an infinite number of enterprises, and initiating and directing a multitude of labors and works which cannot help but provoke difficulties and stir up disputes... and together with this immense and extremely complicated task, it has taken in hand the overall administration and vigilance of the nitrate properties of the state...
>
> This accumulation of enterprises, surpassing perhaps that which exists in any country of Europe or America, demands a vast administrative labor and rigorously requires certificates and contracts of the most varied nature. The government in Chile finds itself today in the situation... of an industrial corporation with an all-embracing line of business, and no matter how great may be its discretion, ... it will not be possible to avoid contentions and disputes...
>
> Our public works affect not only citizens, but also foreign contractors whose capital, industries and labor we are attracting with incessant diligence, and certainly, since it will not be possible to avoid they're suggesting doubts about the meaning of the contracts, or raising difficulties concerning the handling or execution of the works they will both assume the nature of contentions... [65]

Montt's purpose in writing the opinion was to reassert the competency of the judiciary to settle such questions, a competency challenged by Balmaceda in the Nitrate Railways case. But in doing so, he insightfully and even prophetically summarized the dilemma of the Chilean government in the first decade of the Nitrate Age. The state was indeed taking on the characteristics of a great joint-stock company, and factions of the elite were contending for their share of the dividends. Although Balmaceda's economic programs did not inherently threaten the general interests of the elite, they raised the danger of serving one faction at the expense of others. The resulting disputes took on a heightened significance, for the state was no longer simply an arena in which vibrant economic power bases like mining and agriculture were defended. The state itself had emerged as the single most important domestic link to nitrate wealth and possessed the means to affect the beneficial relationships between the nitrate industry and various sectors of the Chilean economy. The government could also provide competitive advantages to rival foreign nitrate producers. Thus, the state became a crucial economic power base from which elite factions and foreign investors sought to benefit.

As the case of the Chañaral railway and the debate on banking policy illustrate, Balmaceda found his plans frustrated by the contest between competing interest groups even among his supporters. The same problem afflicted his attempts to institute a coherent nitrate policy. Fears that it would disrupt the banking sector and that the Edwards clan would be the principal beneficiaries frustrated his initial plan for a national nitrate company. Threats to the banking and agricultural sectors in the form of foreign nitrate service enterprises finally made a project for national companies an acceptable prospect. But when Balmaceda attacked the railway monopoly to lay the groundwork for such companies and ensure increased nitrate revenues, he plunged into a dispute between special-interest groups linked to foreign nitrate producers. Only when he relied solely on his own executive powers to break the railway monopoly and frustrate the British nitrate trust did his policies enjoy even limited success.

His actions indicated his growing belief that only the executive, freed of the contending special-interest groups represented in congress, could carry out effective national policies. Balmaceda's ability to pursue such a course of action was facilitated by the growing resources at the command of the executive. But, as he raised obscure political figures to national prominence through presidential patronage, there emerged a strong suspicion among opposition parliamentarians that Balmaceda was in the process of creating a presidential party in congress to tighten the executive's grip upon the state.[66] Under these circumstances the question of presidential versus congressional authority took on new meaning. The elite could never countenance this enhancement of presidential power now that the state had become such an important economic power base. That such power was being used to benefit select groups within the elite finally made the overthrow of Balmaceda and the presidential system inevitable. On January 6, 1891, the congressional opposition rose in rebellion, hurling the nation into the maelstrom of civil war.

In the Civil War of 1891 the Congressionalists fought beneath the banner of liberty and against the tyranny of the executive. The Balmacedists struggled on behalf of presidential government and against a parliamentary system that gave sway to the personal ambitions of political leaders. But as Marx noted, in historical conflicts it is necessary to "distinguish the phrases and fancies of the parties from their real organism and their real interests, their conception of themselves with their reality."[67] That reality of rule by local elites manifested itself in the de facto parliamentary regime that ruled Chile from 1891 to 1924.

The most significant legislative change after the Civil War was the passage of the municipal reform law in December 1891. This law placed control of the electoral process in the hands of the local municipalities. The immediate political effect of the law was that it "gave the economic elite tremendous control over the registration of voters and supervision of elections."[68] Furthermore, "With the demise of executive authority, not only were local governments to control suffrage, they were also to maximize their own extractive and distributive capabilities ... they gained control of suffrage and direct access to the central coffers of the state."[69]

At the national level these developments reflected the elite's imposition of congressional domination on the governmental process. Servicing political clients through public employment was one overt feature of this new system. The size of the civil bureaucracy ballooned from 5,000 in 1891 to 27,000 by 1919.[70] But, "Of greater significance than the legislator's role in providing employment was his role in providing local brokers with resources from the public treasury for local projects and services. Indeed pork barrel legislation and log rolling to obtain it became the central activities of the Chilean political system."[71] The budget became a shopping list to which congressmen added funding for special projects, while the executive acquiesced in the process to maintain a legislative majority. The state was now characterized by revolving-door ministries and legislative political alliances whose permanence was often measured in days or weeks.

In its dealings with foreign nitrate producers, the congressional regimes pursued the mixed approach that emerged under Balmaceda. His aborted plan to sell state nitrate properties was revived, and five such auctions took place between 1894 and 1903. Chileans participated in the auctions, but foreign producers were the principal beneficiaries. British investment in the industry surged upward from £10 million in 1890 to a peak of £13 million by 1896.[72] Chilean control of the industry did reach 22 percent in 1906 (see Table 16). This figure, however, is probably misleading, since many Chilean corporations formed at that time were the products of a stock-market bubble that burst in 1907.[73] What is significant is the growth in joint Chilean-European ventures (see Table 16). This suggests an expansion of the process that emerged in the Antofagasta Company and in the industrial sector where Chilean interests merged with foreign capitalists to overcome discontinuities in the domestic economy.[74]

Table 16
Percentage of Nitrate Industry Controlled by Chilean and European Investors

Nationality	Year		
	1901	1906	1925
Chilean	15%	22%	14%
Chilean & European	n/a	18	35
British	55	60	28
German	14		13
Other	16		10

Sources: Joseph Robert Brown, "The Chilean Nitrate Industry in the Nineteenth Century" (Ph.D. diss., Louisiana State University, 1954), pp. 120–21; Chile, Ministerio de Hacienda Sección Salitre, Antecedentes, p. 9.

In terms of the industry's service sector, the Nitrate Railways Company's monopoly crumbled as the government granted additional railway concessions. Public criticism intensified against the Bank of Tarapacá and against the Second Combination, which was established by nitrate producers in 1891 to restrict production. They were accused of causing reduced exports of nitrates and bypassing the port of Valparaíso by shipping cargoes directly to Europe. But at the same time, foreign interests still utilized influential Chileans to further their interests in congress. The Mac-Iver brothers continued to promote North's enterprises, and two leading members of the Conservative Party backed the Agua Santa Company.[75] Thus, the elite persisted in its efforts to limit foreign capital where it threatened state revenues and the domestic economy's ties to the industry, while elements of the elite promoted and joined in the ventures of foreign nitrate producers. The Parliamentary Period, then, institutionalized and expanded the symbiotic relationship between foreign capital in the nitrate industry and the Chilean elite.

The destruction of the Chilean presidential system in 1891 and the parliamentary regime that succeeded it were the political consequences that flowed from the resolution of the economic crisis of 1878 when Chile's export markets were captured by more efficient producers. Acceptance of the European nitrate monopoly revived a labor repressive socioeconomic order that had reached its developmental limits. Direct incorporation of the capitalist mode of production with the absorption of the nitrate industry permitted preservation of Chile's traditional society but made the state the focal point of new tensions. Growth in state revenues resuscitated executive domination of the political process that was gradually being eroded before 1879. That resuscitation proved unacceptable owing to the critical economic role now thrust upon the state. Characterized by domestic control of the means of production prior to 1879, Chile underwent a radical shift as the elite abdicated control of the nation's most important resource and its means of production. This abdication left the state as the critical link between the wealth of the nitrate region and the domestic economy. Government coffers became the most important source of domestic nitrate wealth, and state policies could effect the links of the domestic economy to the nitrate industry as well as the competitive struggle among foreign producers. Competition among nitrate producers generated additional domestic political complications as British monopoly capitalism penetrated the state in an effort to protect its fragile position, and the old private producers responded in kind.

The Civil War of 1891 resolved the contradictions between the ever increasing economic importance of the state and the enhanced ability of the executive to channel those resources to select elements of the elite and mediate competition between foreign capitalists. The conflict of 1891 freed elite factions to contend among themselves in the political arena for a share in government wealth and promote the individual interests of foreign nitrate producers. Despite its political bickering and ineffectual policymaking, the parliamentary system was admirably suited to carry out this distributive function. As Marx said of the divided French bourgeoisie's acceptance of parliamentary government, it allowed them to unite "and therefore put the rule of their class instead of the regime of a privileged section of it on the order of the day."[76]

[9]

Conclusion

A NOTED Chilean economist has described the development of the British nitrate monopoly in the 1870s and 1880s as a phenomenon "which at first glance has neither rhyme or reason!"[1] Such an observation is understandable in the light of the gigantic strides taken by the Chilean economy in the first six decades of independence. Yet the Portalian Republic's record of success masked severe internal constraints and elements of dependency. These factors precluded the option of autonomous growth at this juncture and, instead, launched the nation on a new level of dependent development.

Chile's successful experiment in nationhood was the product of precapitalist class relations and commercial exchanges with Europe. In agriculture, the power base of the elite, which shaped the institutions of Chilean society, labor repressive control mechanisms remained predominant. Wages played only a limited role in the lord-peasant relationship. Control of the peon relied far more heavily on the forces of religion, hacendado paternalism, peasant isolation, and a massive labor surplus that served as a powerful negative incentive to the peasantry. Since labor had not emerged as a commodity available only within a competitive market, there was little incentive to reduce labor costs through innovation. Europe, entering the mature stages of industrial capitalism, required raw materials and foodstuffs; but, with steady improvements in productivity in the capitalist center, these requirements did not necessitate a total restructuring of the peripheral societies that supplied them. This was especially true in the Chilean case where the institution of *inquilinaje* and the existence of an amorphous surplus rural population facilitated continued growth in agricultural production through increased exactions upon labor, that is, the production of absolute surplus value. These same conditions obviated tight restrictions on the movement of the labor force, so essential to the maintenance of slave and serf systems. The agrarian order could easily accommodate the limited labor demands of mining. The lines of conflict between city and countryside were less sharply drawn in Chile than in other traditional societies, with their more inflexible forms of lord-peasant relations. Thus, Chilean economic growth was to a significant degree self-generating in sharp contrast to the massive infusions of foreign factors of production during the Guano Age in Peru's more rigidly structured society. But beneath its apparent autocentrism lay significant elements of foreign dependence and domestic strictures on development.

The growth of the economy between 1832 and 1873 depended on European merchant houses that dominated Chile's international trade. Moreover, expansion of exports made the nation increasingly vulnerable to fluctuations in the world commodities market. And while direct foreign investment remained minimal, European inputs, including mining equipment and skilled labor, assured continued growth. This last element indicated the limited innovation capacity of the domestic economy. Foreign production factors were essential to overcome obstacles raised by a society based on nonwage social productive relations. Chilean society had spawned financial, educational, and political institutions geared to the preservation and enhancement of a traditional countryside. These obstacles to development most clearly manifested themselves in the failure of the Chilean nitrate corporations.

The nitrate companies of the 1870s were the product of Chilean economic growth after independence and of the Peruvian economy's structural rigidities. Owing to the inadequacies of the Peruvian system, Chile played a major role in supplying the nitrate industry with foodstuffs, labor, and credit, while Valparaíso emerged as the industry's commercial center. These linkages paved the way for direct penetration by Chilean merchant capital after 1869. The operations of these ventures and those of their European competitors symptomatized the distinct courses of development unfolding in Chile and Western Europe. The Chilean corporations were hampered by the relative scarcity of labor, technology, and technical skills, obstacles inherent in a traditional society. These enterprises depended on foreign sources for credit, marketing, technology, skilled labor, and management. Such inputs were essential in an industry that required steady improvements in productivity. Their indirect access to such factors placed the Chileans at a serious disadvantage in their competition with British and German merchant houses. The commission houses, seeking to offset declines in their customary lines of business, served as agents of industrial capitalism, introducing Europe's powerful capital and industrial resources into the nitrate industry. The only Chilean endeavor to enjoy success was the Antofagasta Company, which brought Chilean and European merchant capital into a direct partnership. Victims of their society's limited capacity for innovation and of their inadequate links to the capitalist center, the remaining Chilean corporations began to flounder even before the initiation of the Peruvian expropriation.

The Peruvian state, anchored to the unstable financial base of guano revenue and wracked by political competition for its resources, had to rely on the large European producers to initiate the expropriation. Peru's weaknesses and the merchant houses' command of capital and other essential factors of production assured the Europeans a favored position under the nationalization process. For the Chilean producers any hope of overcoming domestic obstacles to their development, and the burdens of the expropriation, were destroyed when the national economy's problems reached crisis proportions as world prices began their long-term decline.

Undisputed rulers of their own society, the Chilean elite were nonetheless prisoners of that society's class relations. Chilean society's precapitalist class relations restricted its economic capabilities to the increased production of absolute surplus value. In the face of falling prices for exports, revamping social productive relations was the unacceptable alternative to stagnation and retrogression. A system that had produced forty years of incremental economic progress built on a labor repressive production mode, underpinned by the European commercial network and augmented by selective external infusions of production factors, had exhausted its growth potential. This dependency relationship, which confined interaction between the capitalist center and the peripheral unit to essentially market exchanges, had ensured the continued unity of an elite whose cohesion had been forged through familial linkages in a quiescent agrarian society. This relationship, however, could not survive the new compulsion of the center to drive down prices for primary goods that now represented a disproportionate share of its production costs. Yet Chilean society's superiority over the more fragile economic and political institutions of its neighbors enabled it to enter a new stage of dependent development through victory in the War of the Pacific.

Seizure of the nitrate region layed a solution to the crisis at the feet of the oligarchy. But short of revolutionizing social productive relations, domestic development of the industry was out of the question, and ruling classes have never been prone to premeditated acts of suicide. Given the experience of the Antofagasta Company, some type of direct partnership with foreign capital on an industrywide basis might have been feasible. But the depth of Chile's economic crisis, and pressures from Europe—the dominant force in its international trade and finances—precluded that option. Under these conditions, the elite readily accepted foreign control of the industry. Chilean policies opened the way for the accelerated introduction of European capitalism and its wage labor mode of production.

Unshackled from the restrictions of the Peruvian expropriation, the capitalist Prometheus raised the industry to new levels of productivity and output capacity through capital concentration that rapidly centered control in the hands of a half dozen European enterprises. This direct implant of the capitalist mode of production did not wreak havoc upon Chile's traditional society. There developed rather an articulation or symbiotic relationship between the two modes of production, geographically differentiated as the northern desert and the Central Valley.[2] European capitalism in the north provided fixed capital, technology, skilled labor, management, and marketing, while the Central Valley supplied short-term credit, foodstuffs, and unskilled labor. Contradictions within the articulation, particularly limits on the effectiveness of the wage labor market, and the inefficiency of agriculture were overcome through capitalist productivity and the policies of the state. European capitalism enjoyed easy access to a vital resource and the freedom to implant its mode of production to a greater degree than ever before possible. For the Chilean elite, linkages to the north in the form of taxation and market relations resuscitated the faltering domestic economy. The Central Valley was now shielded from the secular decline in world prices as its market relations shifted from an external focus to the capitalist production of the northern desert.

While experiencing new levels of economic growth, Chile's social and economic structures grounded in the premodern class relations of the countryside survived virtually unchanged. The viability of coopting dynamic urban elements into the elite was also preserved. The process was enhanced by increasing infusions of European factors of production in the Central Valley that muted potential conflict between industry and agriculture, bypassing but not undermining internal obstacles to industrialization. Therefore, Chile followed neither the pattern of England, where the capitalist mode of production in agriculture and industry assured complementarity of the two sectors, not that of Eastern Europe, where serfdom suffocated incipient industrialization.[3] Links to the nitrate industry and externally produced factors of production permitted the oligarchy to straddle the apparently contradictory power bases of an archaic countryside and modern industry. Even late arrivals to elite status received practical rewards from their position as members of the landed oligarchy. These included social standing, improved access to the national credit market, and most importantly political power, which assumed an economic significance of its own in the new dependency relationship that underlay the Nitrate Age.

Fernando H. Cardoso has distinguished two basic types of dependency relationships. In the first, the elite retain control of the nation's natural resources and means of production, in the second, the bourgeoisie relies on taxation of foreign production of those resources.[4] With the onset of the Nitrate Age, Chile achieved a successful transition from the first type to the second. The transformation, however, was not complete, since the elite still controlled significant segments of the national economy. This fact, and the preexisting coherence of the elite, prevented the shattering of the bourgeoisie so often predicated as an effect of dependency.[5] Yet the transition did create disruptions including the rise of the working class and middle-income groups that the oligarchy would eventually have to repress or coopt. Of more immediate concern were the tensions building in the state, the linchpin in the linkages between European capitalism in the nitrate zone and Chile's traditional society.

As the first decade of the Nitrate Age came to a close, the ownership structure of the industry was altered by the forces of monopoly capitalism. With raw materials produced by labor-intensive methods on the periphery accounting for a disproportionate share of its production costs, the capitalist center sought to drive down their value. In the case of the nitrate industry, European merchant capital had spearheaded this effort, ensuring ready access to the means of production for British monopoly capital. Between 1888 and 1889 many of the large *oficinas* passed under the control of British joint-stock companies. Efforts to protect corporations that did not function as strictly competitive capitalist enterprises led to direct penetration of the Chilean political process. In response, private companies duplicated this manuever, seeking to turn the now significant influence of the state to their competitive advantage. In Chile the state had emerged as the oligarchy's principal means of access to nitrate wealth. A reinvigorated presidency, however, threatened to channel that wealth and power over the nitrate industry to select elements of the elite. The result was an acute political crisis as the state reached beyond its customary function of moderator of elite economic activity to that of mediator between the capitalist north and the traditional society of the Central Valley. These tensions shattered the presidential system in the Civil War of 1891 and gave birth to a parliamentary regime that could more readily accommodate politicized economic competition among the elite and their representation of individual foreign interests.

During the Parliamentary Republic (1891–1924), the new relationship with the capitalist center stabilized and the tensions and pressures that marred the first decade of the Nitrate Age subsided. Under these conditions, foreign investors continued to dominate the nitrate industry. Chileans successfully followed the precedent of the Antofagasta Company, forming joint nitrate ventures with European capitalists to bypass the structural inadequacies of the domestic economy. Meanwhile, the oligarchy used the parliamentary process to compete for state nitrate revenues, promote the interests of European nitrate producers, and limit threats to the market linkages between the Central Valley and the nitrate region. The Parliamentary Period's chaotic politics represented the resolution of contradictions in a state coping with the heterogeneous structures arising from incorporation of capitalist production into a precapitalist society. Chile's ruling class had successfully made the transition into a new dependency relationship with the capitalist center. But like the earlier relationship, it would fail to generate autocentric capitalist development.

The nitrate industry catapulted Chile into an age of dazzling prosperity. Yet beneath its gilded surface, Chilean society remained stalled in a process of underdeveloped or peripheral capitalism. Nitrate wealth and the quickening pace of industrialization failed to prompt demands for a radical restructuring of regressive institutions such as agriculture and education. This absence of progressive reform is readily apparent in the relationship between agriculture and manufacturing. Labor repressive mechanisms of control and low productivity placed severe limits on the domesitc consumer market and inflationary pressure on food prices. Yet industrialists launched no attack on agriculture or the tariff policies that protected it.[6] That many industrialists sprang from or were coopted into the landed oligarchy is further evidence of the compatibility of what should have been contradictory interests.[7] This failure of the Nitrate Age to spawn social forces that would press forward in restructuring Chilean society was a product of the new dependent relationship established with European capitalism.

The direct penetration of European capitalism in the nitrate industry and manufacturing enabled Chile to maintain its position in the world economy and undertake a limited process of industrialization. Tied to European capitalism through direct partnerships, state taxation, and sec-

CONCLUSION

toral linkages, the elite enjoyed continued economic growth while their agrarian power base was preserved intact. This alliance with the capitalist center overcame the seemingly inevitable conflict between enterprises requiring productivity improvement and a traditional socioeconomic system. It thereby prevented the formation of a consensus among the elite for progressive reforms to achieve mature self-generating capitalism, while it left the oligarchy dependent on continued inputs from the capitalist center to maintain this new order.[8]

The possibility for alternative courses of action was further limited by the condition of Chile's nonelite groups. Whatever its implications for long-term development, the nitrate industry created sufficient growth in the domestic economy to satisfy local merchants and other business interests. Although the achievement of elite social status remained a formidable and complex task for such groups, the possibility of reaching social prominence remained a real one. The resolution of the potential conflict between elite agricultural and industrial interests, and continued opportunities for social and economic advancement for the business community, precluded pressure for change from upper-income groups. The potential for reformist movements from lower down the social ladder was inhibited by the structures of Chilean society that predated the dawn of the Nitrate Age. It would be difficult to prove the existence of a middle class at this time, much less assess its potential as a force for change. A labor movement had surfaced in the port cities and the nitrate region. But organizational efforts remained isolated in those areas, since most of the urban work force was scattered among numerous artisanal enterprises or struggled to survive through a series of occasional employments.[9] As for the peasantry, effective elite mechanisms of control and the absence of villages as centers of resistance prevented them from offering meaningful challenges to their hacendado masters. The inability of the lower social orders to organize and achieve such reforms as effective limits on the working day and increases in real wages precluded yet another source of pressure for rapid improvement in the domestic economy's efficiency. The disorganization of the lower classes sealed off the last possible source of change within Chilean society, permitting the elite to pursue policies that protected and promoted the nondynamic structures on which they built their power.

Clearly, the potential for autocentric Chilean development was extremely limited. The rule of a coherent elite grounded in nonwage social relations and the disorganization of the lower social classes offered scant opportunities for class alliances that would press for structural change. If the decline of the export economy had not been halted by the absorption of the capitalist mode of production in the form of the nitrate industry, the prospects for Chilean development would still have been bleak. It is possible that Chile would have experienced continued economic decline and a tightening of labor repressive methods of production; and its society might have gradually slipped into a condition of autarky. Yet such a hypothesis banishes one half of the relationship that explains the evolution of Chilean society. For while internal class relations help define the potential of Chilean history, their actual role can be fully understood only in their interaction with the forces emanating from the capitalist center.

One might infer from recent writings on dependency that attempts to fathom the causes of underdevelopment confront an either-or proposition. Underdevelopment is either caused by internal social productive relations generated by the class struggle or by market relations characterized by unequal exchange.[10] Chile's experience suggests instead a complementary relationship between the two. Class relations do set limits of development, but as these limits are reached the capitalist center can spur the system to new levels of growth without totally transforming class relations even after the direct infusion of the capitalist mode of production. In Chile, the dependent relationship was not based on capitalist exploitation of readily available surplus labor. Instead, it centered on the capacity of capitalist productivity to overcome without destroying labor scarcity and other attendant consequences of nonwage class relations. This symbiotic relationship opened the way for the periphery to experience continued economic growth while preserving precapitalist class relations. In return, the capitalist center enjoyed the protection of the domestic power structure in further implementing its mode of production in order to reduce the value of imported raw materials. Inevitably, this order reached its limits and had to be transcended.

In Chile the Nitrate Age set the pattern of development for the first half of the twentieth century with United States capital in copper eventually replacing British capital in nitrates as the growth pole of the economy. Extension of state intervention to maintain the heterogeneous and often contradictory segments of the national economy spawned, within the public bureaucracy, a quasi-middle class that had little concern with the creation of an independent economic base. Continued penetration of foreign capital only exacerbated the fragmentation of the working class. These developments continued to diffuse internal pressures for structural change. The state's pivotal position as mediator between the capitalist and precapitalist segments of society resulted in continued political crises as the elite and new social groups struggled to maintain or enhance their positions primarily through the political process. As elite control of political institutions was challenged, the oligarchy resorted to tacit alliances with middle-income and labor groups to preserve the agrarian order while national business associations helped maintain its influence on state economic policies.[11] Now that this relationship has reached its limits and import substitution has failed to provide a satisfactory alternative, Chile's ruling class is seeking a symbiotic relationship with multinational capitalism. With a decline in raw material prices that began in the early 1950s, multinationals shifted their focus in the periphery to the production of finished goods. Attempting to accommodate this new order and break the latest cycle of internal stagnation, the current military regime has opened the floodgates to a new wave of foreign investment. In the name of free enterprise, the way has been cleared for further penetration of the most dynamic sectors of the economy by multinational capitalism. Some elite-controlled portions of the economy are being sacrificed, and those social groups that are not linked to the international sector are increasingly marginalized. Owing to these destabilizing effects of the new order, mechanisms of control such as paternalism, segmentation, and limited cooptation have been supplemented by institutionalized violence and fear. But this latest transition in the dependency relationship seems no more likely than those of the past to create the conditions for autocentric development.

Appendix

Labor Productivity in Tarapacá, 1880–90

Year	Production (Quintals)	Workers	Quintals per Worker
1880	4,869,000	2,848	1,709
1881	7,739,000	4,906	1,577
1882	10,701,000	7,124	1,502
1883	12,820,000	7,077	1,811
1884	12,152,000	6,505	1,868
1885	9,478,000	4,571	2,073
1886	9,805,000	4,534	2,162
1887	15,300,000	7,201	2,124
1888	16,682,000	9,180	1,816
1889	20,681,000	11,422	1,810
1890	22,397,000	11,657	1,921

Sources: Chile, Ministerio de Hacienda, *Memoria de 1890*, p. 81; *Memoria de 1890–1891*, p. 251.

Wages and Food Prices (Current Pesos) for Nitrate Workers and Peons, 1871–90

	Nitrate Worker				Peon[2]		
Years	Average Daily Wage	Flour 46 kg	Beans 100 kg	Years	Average Daily Wage	Flour 46 kg	Beans 100 kg
1873	.90	—	—	1871–75	.27	2.95	4.29
1881		4.90	6.00	1881		3.92	3.20
1882		3.84	5.91	1882		3.20	3.37
1883		4.03	9.05	1883	.30	3.20	4.00
1884		3.55	14.45	1884		3.32	8.50
1885		4.62	8.19	1885		3.85	5.12
1886	1.37	4.23	9.39	1886		3.45	3.65
1887		4.66	8.94	1887		3.70	5.20
1888		4.50	9.76	1888	.30	4.18	6.90
1889	6.00	6.64	12.51	1889		5.27	6.95
1890	4.00	6.20	11.75	1890		4.87	6.00

Sources:
1. Wages: Bohl to Hayne, Valparaíso, 16 May 1873. GMS, 11, 121; Brown, "Nitrate Industry," p. 197. Foods: Beans and flour were the staples of the workers' diet. Prices are for Iquique; see El Véintiuno de Mayo (Iquique), 1881–88, La Industria (Iquique), 1882–90.
2. Bauer, Rural Society, pp. 156, 233–34.

The usual qualifications must be made in assessing these statistics. Nitrate workers no doubt paid higher prices for food than the figures would indicate, since they expended 60 percent of their wages purchasing necessities from company stores. Peons, on the other hand, were shielded to some degree from price increases by nonwage payments and subsistence plots. But even if we assume that prices at the company stores were twice as high as those prevailing in the Central Valley, the fact remains that nitrate workers regularly earned four times as much as agricultural laborers.[1] The drastic difference in both nominal and real wages, aided by underemployment in the countryside, apparently produced a rural to urban migration. The population of the northern mining region and major urban centers (over 10,000 inhabitants) increased by 122 percent between 1865 and 1895, while the total national population rose by only 48 percent. Yet despite this migration, both nitrate producers and mineowners constantly encounterd problems in maintaining an adequate work force.

Labor shortages are, of course, relative. Ideally, nitrate producers, mineowners, and agriculturists all sought a surplus of labor large enough to prevent any increase in real wages, and wages in Chile were generally low by the standards of the industrialized nations. But only landowners appear to have succeeded in stabilizing the cost of labor. The size of the rural population remained stable, and real wages in the countryside stagnated or fell.[2] Conversely, nitrate and mining operations, despite high wage levels, competed for the same limited pool of labor. In the early 1870s the Antofagasta Company reported difficulty in lowering real wages due to competition from nearby mines, and producers in Tarapacá bemoaned the scarcity of workers throughout the decade.[3] In turn, mining interests from Copiapó advertised in Iquique in 1888 offering jobs to 2,000 workers, and in 1889 the British consul in Coquimbo reported a labor shortage owing to "the heavy losses from cholera, and the great emigration of the working classes to the nitrate districts in the north, where they get better pay."[4] Thus, in Chile nonwage-control mechanisms continued to anchor a rural population sufficient to meet the needs of a labor-intensive agrarian system, while enterprises relying on wages found themselves competing against each other in a closely circumscribed market.

A number of reasons can be offered for the apparent limitations on the wage labor market. Rural isolation that impeded communications, abysmal or nonexistent educational institutions that caused a shortage of skilled workers, the subhuman working conditions in the nitrate deserts, the preference of *salitreros* and mineowners to draw from each other's experienced labor forces – all can be granted some validity as causal factors. Yet they are as much symptoms as they are causes; for it is apparent that Chile at this time was characterized by dual labor markets, distinct in terms of their methods of control. It is equally apparent that the social productive relationships of the countryside not only succeeded in retaining the necessary labor force but shaped, or rather distorted, other societal institutions (e.g., education) to maintain the existing order.[5] Modern or wage sectors were then left to compete for a scarce labor supply and compensate for relative labor scarcity with productivity improvement. And since the innovations were achieved through the introduction of foreign factors of production, fundamental challenges to Chile's traditional society could be avoided.

While it is often assumed that peripheral societies contain vast armies of surplus labor easily exploited by the capitalist center, Chile at this stage of its development hardly supports such a hypothesis. Instead, capitalist enterprises relied on productivity improvement to cope with labor scarcity resulting from the premodern class relations of the countryside, and the external generation of those improvements facilitated preservation of Chile's traditional society. It was this ability of capitalism – to generate cost-effective production of nitrate for the world market without disrupting the traditional agrarian sector – that formed the basis of the symbiotic relationship between European nitrate producers and the Chilean elite.

APPENDIX

British Control of Nitrate Production in 1890

Oficinas	Nitrate Sent to Port in Tarapacá in 1890 Amount (metric quintals)[a]	Percentage
23 British Oficinas	7,115,496	69%
25 Non-British Oficinas	3,217,148	31%
Total:	10,332,664	100%

Total nitrate exported from Tarapacá in 1890: 9,895,892.13

9,875,892.13 metric quintals x 69% = 6,828,165.56 metric qtls.

6,828,165.56 + 574,551.08 metric quintals exported from Taltal[b] = 7,402,716.64 (Total British export of nitrate from Chile)

   7,402,716.64 metric quintals (Total British export)
÷ 10,758,904.04 metric quintals (Total exported from Chile, i.e., Tarapacá, Antofagasta, and Taltal)
= 69% – Percentage of total Chilean nitrate exports produced by British-owned oficinas in 1890

Source: Chile, Ministerio de Hacienda, Memoria de 1892 (Santiago, 1892), pp. 249–53.
[a]One metric quintal = 220.42 pounds.
[b]Only British-owned oficinas were operating in Taltal at this time; see Chile, Ministerio de Hacienda, Memoria de 1890, p. 75.

Exchange Rate for Chilean Pesos and Peruvian Soles (in pence of a British pound, rounded to nearest d), 1869–90

Year	Pesos[1]	Soles[2]
1869	46	48
1870	46	49
1871	46	48
1872	46	44
1873	45	—
1874	45	—
1875	44	41
1876	41	28
1877	42	21
1878	40	30
1879	33	—
1880	31	—
1881	31	—
1882	35	—
1883	35	—
1884	32	—
1885	26	—
1886	24	—
1887	24	—
1888	26	—
1889	27	—
1890	24	—

Sources:
1. Frank W. Fetter, Monetary Inflation in Chile (Princeton, 1931), p. 13.
2. Figures have been derived from a number of sources, including the Gibbs manuscripts and British consular reports. Given the wild fluctuations that Peruvian currency underwent, these figures are at best rough approximates.

Notes

NOTE ABBREVIATIONS

Archival Collections

Santiago

FN	Fondo Nuevo "Varios," National Archive
JI	Archivos Judiciales de Iquique, National Archive
MH	Archivo del Ministerio de Hacienda, National Archive
MRE	Archivo del Ministerio de Relaciones Esteriores, National Archive
NI	Archivos Notariales de Iquique, National Archive
NS	Archivos Notariales de Santiago, National Archive
NT	Archivos Notariales de Tarapacá, National Archive
NV	Archivos Notariales de Valparaíso, National Archive

London

F.O. 16	Foreign Office Archives, General Correspondence—Chile, Public Record Office
F.O. 61	Foreign Office Archives, General Correspondence—Peru, Public Record Office
F.O. 177	Foreign Office Archives, Embassy and Consular Archives, Correspondence—Peru, Public Record Office
GMS	Archives of Anthony Gibbs and Sons, Guildhall Library

Government Publications

BSES (year)	Chile, Congreso, *Boletín de las sesiones estraordinarias de la cámara de senadores*
BSOD (year)	Chile, Congreso, *Boletín de las sesiones ordinarias de la cámara de diputados*
BSOS (year)	Chile, Congreso, *Boletín de las sesiones ordinarias de la cámara de senadores*
PP (year)	Great Britain, Parliament, *Parliamentary Papers*

Periodicals and Journals

BELC *Boletín de Estudios Latinoamericanos y del Caribe*
BOI *Burdetts Official Intelligence*
FN (London) *Financial News*
HAHR *Hispanic American Historical Review*
JLAS *Journal of Latin American Studies*
NLR *New Left Review*
PHR *Pacific Historical Review*
SAJ *South American Journal*
TA *The Americas*

CHAPTER I

1. Celso Furtado, *Economic Development of Latin America: A Survey From Colonial Times to the Cuban Revolution*, trans. Suzette Macedo (Cambridge, England, 1970), p. 24.
2. Alberto Herrmann, *La producción de oro, plata i cobre en Chile desde los primeros dias de la conquista hasta fines de agosto de 1894* (Santiago, 1894), pp. 43, 45–48; Octavio Astorquiza, comp. and ed., *Lota: antecedentes historicos, con una monografía de la Compañía Minera e Industrial de Chile* (Concepción 1929), p. 134.
3. Arnold J. Bauer, *Chilean Rural Society from the Spanish Conquest to 1930* (Cambridge, England, 1975), pp. 62–65. On peso exchange rates, see Appendix.
4. Ramón E. Santelices, *Los bancos chilenos* (Santiago, 1893), pp. 56–84, 183.
5. Luis Escobar Cerda, *El mercado de valores* (Santiago, 1959), p. 52.
6. Bauer, *Rural Society*, p. 30.
7. Ibid., p. 94.
8. Harold Blakemore, *British Nitrates and Chilean Politics, 1886–1896: Balmaceda and North* (London, 1974), pp. 11–14; Claudio Véliz, *Historia de la marina mercante de Chile* (Santiago, 1961), pp. 79–183; Bauer, *Rural Society*, p. 27.
9. Markos J. Mamalakis, *The Growth and Structure of the Chilean Economy: From Independence to Allende* (New Haven, Conn., 1976), p. 19.
10. Daniel Martner, *Estudio de la política comercial chilena e historia económica nacional*, 2 vols. (Santiago, 1923), vol. I, pp. 275–76.
11. Arturo Valenzuela, *Political Brokers in Chile: Local Government in a Centralized Polity* (Durham, N.C., 1977) pp. 184–85; Germán Urzúa Valenzuela, *Evolución de la administración pública chilena (1818–1968)* (Santiago, 1970), pp. 80–93.
12. Francisco A. Encina, *Historia de Chile desde la prehistoria hasta 1891*, 4th ed., 20 vols. (Santiago, 1955), Vol. XIII, pp. 585–601.
13. E. Semper and E. Michels, *La industria del salitre en Chile*, trans. and augmented by Javier Gandarillas and Orlando Ghigliotto Salas (Santiago, 1908), pp. 7–18.
14. Roberto Hernández Cornejo, *El salitre: resumen histórico desde su descubrimiento y explotación* (Valparaíso, 1930), p. 174; Oscar Bermúdez Miral, *Historia del salitre desde sus orígenes hasta la Guerra del Pacífico* (Santiago, 1963), pp. 154–58.

15. Hernández, *Salitre*, p. 46; Bermúdez, *Historia del salitre*, p. 150.
16. Jonathan V. Levin, *The Export Economies* (Cambridge, Mass., 1960), pp. 34–44.
17. Enrique Kaempffer, *La industria del salitre: Anexos y glosario* (Santiago, 1914), pp. 152–54; Bermúdez, *Historia del salitre*, p. 137.
18. Bermúdez, *Historia del salitre*, pp. 56–62.
19. Hernández, *Salitre*, p. 37; Guillermo E. Billinghurst, *Los capitales salitreros de Tarapacá* (Santiago, 1889), pp. 36–37.
20. *South Pacific Times* (Callao), 14 September 1872. On exchange rates for the Peruvian sol, see Appendix.
21. John Smail to Alfred Bohl, Oficina Limeña, 13 June 1878. GMS, 11,472/1. Smail, Bohl, James Charles Hayne, James Henry, Thomas Denison Comber, Brice Alan Miller, Henry Manuel Read, Henry Giles Daubeny, John Manely Lowe, and Alfred Naylor were all partners in William Gibbs and Company (known as Gibbs and Company after 1879), the west coast branch of Antony Gibbs and Sons; see W. Maude, *Antony Gibbs & Sons Limited: Merchants and Bankers, 1808–1958* (London, 1958), p. 130.
22. Bauer, *Rural Society*, pp. 52–53.
23. Hernández, *Salitre*, pp. 27–28, 45, 70; Billinghurst, *Capitales salitreros*, pp. 34, 36.
24. Hernández, *Salitre*, pp. 46, 54.
25. Robert G. Greenhill and Rory M. Miller, "The Peruvian Government and the Nitrate Trade, 1873–1879," *JLAS* 5 (May 1973), p. 112; "Memorandum as to the Nitrate Business of Peru," n.p., n.d. GMS 11,132.
26. Bermúdez, *Historia del salitre*, pp. 247–48.
27. *El Mercurio* (Valparaíso), 30 June 1872; Emilio Romero, *Historia económica del Perú*, 2d ed., 2 vols. (Lima, 1968), vol. 2, p. 146; *South Pacific Times* (Callao), 21 May 1872.
28. *South Pacific Times* (Callao), 14 September 1872.
29. Bermúdez, *Historia del salitre*, pp. 189–233.
30. Billinghurst, *Capitales salitreros*, pp. 37–38.
31. *NI*, vol. 7, 2 July, 15 July, 17 July 1872, no f. or doc. nos.; Compañía Salitrera Valparaíso, *Estatutos* (Valparaíso, 1872), pp. 3–6.
32. Compañía Salitrera San Carlos, *Estatutos* (Valparaíso, 1873), pp. 3–4; *NI*, vol. 9, 15 January 1873, no f. or doc. nos.
33. Virgilio Figueroa, *Diccionario histórico y biográfico de Chile*, 5 vols. (Santiago, 1926–35), vol. II, p. 553, vol. V, p. 593.
34. Compañía Salitrera Valparaíso, *Estatutos*, pp. 3–4; Compañía Salitrera America, *Estatutos* (Valparaíso, 1873), p. 3; Compañía Salitrera Sacramento, *Estatutos* (Valparaíso, 1872), pp. 3–4; Compañía Salitrera Negreiros, *Estatutos* (Valparaíso, 1872), pp. 3–4; *NV*, vol. 171, 25 April 1872, fs. 461–65.
35. Bauer, *Rural Society*, p. 54.
36. Bauer, *Rural Society*, pp. 101, 122, 150, 155–59, 163–67. On the development implications of labor repressive agrarian systems, see Robert Brenner, "The Origins of Capitalist Development: A Critique of Neo-Smithian Marxism,"

NLR, 104 (July–August 1977), pp. 47–51, 68–73.
37. A number of mercantile partnerships recorded in the Archivos Notariales de Valparaíso at this time specifically excluded credit extensions to nitrate *oficinas* from their sphere of business; see, for example, *NV,* vol. 179, 1 December 1873, fs. 875–78. On the structural inadequacies of the national credit market, industrial base, and educational institutions, see Henry W. Kirsch, *Industrial Development in a Traditional Society* (Gainesville, Fla., 1977), pp. 14, 58–59; Mamalakis, *Chilean Economy,* pp. 74–78; *La Epoca* (Santiago), 6 June 1884.
38. Gabriel Marcella, "The Structure of Politics in Nineteenth Century Spanish America: the Chilean Oligarchy, 1833–1891" (Ph.D. diss., University of Notre Dame, 1973), pp. 90–126. Marcella documents the extent of familial connections but argues that diversity of economic endeavors indicates that landowners did not dominate; see, however, Bauer's argument in Bauer, *Rural Society,* pp. 215–17.
39. Bauer, *Rural Society,* pp. 155–57. The prevalence and persistence of the problem is captured in a statement by the manager of the most important *oficina* in Tarapacá, that in terms of wages refinery workers "have for years had things pretty much their own way." Smail to Read, Oficina Limeña, 3 February 1879. GMS, 11,472/11. See also Semper and Michels, *Industria del salitre,* pp. 80–81, 100. On the depressing effect that rural conditions had on wage levels in Chile, see Ann Louise Hagerman Johnson, "Internal Migration in Chile to 1920: Its Relationship to the Labor Market, Agricultural Growth and Urbanization," (Ph.D. diss., University of California, Davis, 1978), pp. 270–98.
40. Bohl to Hayne, Valparaíso, 16 May 1873, GMS, 11,121.
41. Semper and Michels, *Industria del salitre,* p. 69; William Howard Russell, *A Visit to Chile and the Nitrate Fields of Tarapacá* (London, 1890), p. 184.
42. Use of European technology included introduction of Marsden roll-and-jaw crushers in 1873, and modifications in the refining process culminating in 1879 with the Shanks system. The system was first used in the British chemical industry. The process was patented in 1863, and the first *oficina* to adapt it to nitrate refining was under construction in 1878. This compares quite favorably with the time required for adoption of technological improvements such as the Solvay process in the European industrial community. M.B. Donald, "History of the Chile Nitrate Industry," *Annals of Science* 1 (January 1936), pp. 44–47; 1 (April 1936), p. 195; David S. Landes, *The Unbound Prometheus* (Cambridge, England, 1969), pp. 271–72.
43. Semper and Michels, *Industria del salitre,* p. 100; *PP,* 1890, vol. 73, J. G. Kennedy to Marquis of Salisbury, Santiago, 20 November 1889, "Report on European Emigration to Chile," p. 7.
44. Billinghurst, *Capitales salitreros,* p. 37.
45. *NV,* vol. 174, 20 December 1872, f. 748; vol. 184, 7 January 1874, f. 11; see also the nitrate company statutes cited in note 34, above.
46. Table by Juan Ibarra, n.p., 31 May 1873. GMS, 11,129.
47. *El Mercurio* (Valparaíso), 13 December 1872.

48. Compañía Salitrera Pisagua, *Memoria* (Valparaíso, 1872), p. 5; *El Mercurio* (Valparaíso), 4 February 1872, 18 August 1872.
49. Heraclio Bonilla, *Guano y burguesía en el Perú* (Lima, 1974), pp. 25–53.
50. Samir Amin, *Unequal Development*, trans. Brian Pearce (New York, 1976), pp. 185–86.
51. Greenhill and Miller, "Nitrate Trade," p. 119. Concerning the declines in their traditional lines of business that led European merchant houses to make and maintain direct investments in enterprises such as nitrates, see D. C. M. Platt, *Latin America and British Trade, 1806–1914* (London, 1972), pp. 136–43.
52. Antony Gibbs & Sons (cited hereafter as AGS) to William Gibbs & Company (cited hereafter as WGC), London, 29 January 1864. GMS, 11,471/1; Annual Accounts of William Gibbs and Company, Valparaíso, 30 April 1857, 30 April 1865. GMS, 11,033/3/5. Billinghurst, *Capitales salitreros,* pp. 31–32.
53. Table by Juan Ibarra, n.p., 31 May 1873. GMS, 11,129; Yearly Accounts of the Tarapacá Nitrate Company, Iquique, 31 December 1881, f. 64. GMS, 11,049 A. Estimates of the Limeña's capacity vary from 600,000 to 1 million quintals; see Bermúdez, *Historia del salitre,* p. 166; Read to Henry, Iquique, 21 August 1873. GMS, 11,123. The present estimate is based on the *oficina's* highest annual production of 781,968.44 quintals in 1877; see Yearly Accounts of the Tarapacá Nitrate Company, Iquique, 31 December 1881, f. 64. GMS, 11,049 A. Since nitrate *oficinas* rarely, if ever, produced at full capacity for an entire year, a slight allowance for this has been made to arrive at the figure of 800,000 quintals.
54. Bohl to Read, Iquique, 17 May 1873. GMS, 11,121; Hayne to Bohl, Valparaíso, 1 May 1873. GMS, 11,120; Annual Accounts of the Tarapacá Nitrate Company, Iquique, 30 April 1875. f. 40. GMS, 11,049 A; AGS to WGC, London, 15 December 1875. GMS, 11,471/2.
55. Bermúdez, *Historia del salitre,* p. 272. Guillermo Billinghurst claimed that the firm was partly Peruvian, but the only Peruvians were the wives of the owners who became the principal shareholders on the deaths of their husbands; see Billinghurst, *Capitales salitreros,* p. 33; *NI,* vol. 68, 3 April 1882, fs. 90–91. On *oficina* productive capacity, see Table by Juan Ibarra, n.p., 31 May 1873. GMS, 11,129.
56. Bermúdez, *Historia del salitre,* pp. 266–67; *NI,* vol. 68, 8 April 1882, fs. 99–103; Table by Juan Ibarra, n.p., 31 May 1873. GMS, 11,129.
57. Clark, Eck and Company to Nairn, Iquique, 6 October 1873. Copy encl. in Jerningham to Granville, Lima, 24 October 1873. F.O. 61/279. Nairn was the British consul in Iquique, Jerningham the British minister in Lima, and Granville the foreign secretary; *NI,* vol. 7, 23 June 1872, no f. or doc. no.; *NV,* vol. 171, 6 June 1872, fs. 658–60.
58. The Gibbs branches and the London office remained in weekly contact by mail and used telegraphic ciphers for urgent messages.
59. *NV,* vol. 179, 1 September 1873, f. 211; *NV,* vol. 184, 10 April 1874, f. 196; Read to Bohl, Oficina Limeña, 12 October 1878. GMS, 11,123; Billinghurst,

Capitales salitreros, p. 104.
60. W. M. F. Castle, *Sketch of the City of Iquique, Chili, South America during Fifty Years* (Plymouth, England, 1887), p. 43. Although written in 1887, Castle's description of the efficiency of Gibbs's nitrate operations holds true for the 1870s and is thoroughly documented in the Gibbs correspondence and account books for the period.
61. Thomas F. O'Brien, "The Antofagasta Company: A Case Study of Peripheral Capitalism," *HAHR* 60 (February 1980), pp. 8–9, 26.
62. On the persistence of traditional production techniques in European agriculture, see E. J. Hobsbawm, *The Age of Capital, 1848–1875* (New York, 1975), pp. 179–81. On problems in using nitrate of soda in mechanized agriculture, see *Scientific American* 61 (1889), p. 407.
63. Miguel Cruchaga, *Salitre y guano* (Madrid, 1929), p. 173.
64. Basil Lubbock, *The Nitrate Clippers* (Glasgow, 1932), pp. 111–15; Cruchaga, *Salitre y guano,* p. 173.
65. Hayne to Bohl, Valparaíso, 26 April 1873. GMS, 11,122.
66. Bonilla, *Guano y burguesía,* pp. 69–93.
67. S. B. Saul, *Studies in British Overseas Trade, 1870–1914* (Liverpool, 1960), pp. 90–101.
68. Charles A. McQueen, *Peruvian Public Finance,* Department of Commerce, Bureau of Foreign and Domestic Commerce, Trade Promotion Series No. 30 (Washington, D.C., 1926), p. 87; Levin, *Export Economies,* pp. 97–104.
69. Levin, *Export Economies,* pp. 105–6; Greenhill and Miller, "Nitrate Trade," pp. 112–13; Romero, *Historia económica,* 2, p. 147; Hernández, *Salitre,* p. 89; Carlos Aldunate Solar, *Leyes, decretos i documentos relativos a salitreras* (Santiago, 1907), part I, pp. 27–32.
70. J. D. Campbell to Bohl, Excerpt of a letter contained in Henry to Hayne, Lima, 9 September 1874. GMS, 11,121.
71. Bermúdez, *Historia del salitre,* p. 323; Read to Henry, Iquique, 31 July 1873, 21 August 1873. GMS, 11,123; Jerningham to Granville, Lima, 27 September 1873. F.O. 61/279.
72. *Valparaíso and West Coast Mail,* 6 September 1873; Read to Henry, Iquique, 11 September 1873. GMS, 11,123.
73. *Valparaíso and West Coast Mail,* 18 January, 14 June, 5 July 1873; Comber to Read, Valparaíso, 16 September 1873. GMS, 11,122; Hayne to Comber, Iquique, 26 February 1875. GMS, 11,128.
74. Guillermo Subercaseaux, *Monetary and Banking Policy of Chile* (Oxford, 1922), p. 191; Rumbold to Granville, Santiago, 16 September 1873. F.O. 16/177.
75. *Brazil and River Plate Mail* (London), 8 May 1873; Read to Henry, La Noria, 2 July 1873. GMS, 11,123; Jerningham to Granville, Lima, 24 February 1873. F.O. 61/279.
76. *La Patria* (Valparaíso), 15 March 1873, 8 May 1874; *NV,* vol. 179, 17 February 1873, f. 507; *Valparaíso and West Coast Mail,* 13 February 1874; Read to Bohl, Iquique, 28 January 1875. GMS, 11,123.

NOTES

77. Comber to Read, Valparaíso, 24 September 1873, 1 October 1873, 16 October 1873. GMS, 11,122.
78. *NV*, vol. 184, 16 March 1874, fs. 222–23.
79. *NV*, vol. 185, 7 January–28 January 1874, fs. 2–19; *NT*, vol.12, Iquique, 4 February, 4 July 1874, no f. or doc. nos.
80. *NV*, vol. 183, 12 September 1874, fs. 432–34; *NV*, vol. 184, 5 May 1874, fs. 400–401, 25 May 1874, fs. 469–70.
81. Read to WGC at Lima, Iquique, 22 June 1874. GMS, 11,123.
82. O'Brien, "Antofagasta Company," pp. 9, 11.
83. Henry to Hayne, Lima, 6 June 1874. GMS, 11,121.
84. Clark, Eck and Company to Nairn, Iquique, 6 October 1873. Copy encl. in Jerningham to Granville, Lima, 24 October 1873. F.O. 61/279; Read to Henry, Iquique, 9 October 1873. GMS, 11,123.
85. Henry to Hayne, Lima, 24/25 January, 29 January, 1 February, 7 February 1873. GMS, 11,121; Hayne to Henry, Valparaíso, 18 January 1873. GMS, 11,120; Read to WGC at Lima, Iquique, 22 June 1874. GMS, 11,123.
86. Henry to Hayne, Lima, 10/12 March 1873. GMS, 11,121; Read to Henry, La Noria, 2 July 1873. GMS, 11,123; Hayne to Henry, Valparaíso, 10 October 1874. GMS, 11,120.
87. Bermúdez, *Historia del salitre*, pp. 267–68.
88. Hayne to Henry, Valparaíso, 11 April, 28 May, 16 June 1874. Hayne to Bohl, Valparaíso, 19 April, 23 April 1873. GMS, 11,120; Bermúdez, *Historia del salitre*, pp. 267, 272.
89. Read to Henry, Iquique, 16 September 1873. GMS, 11,123.
90. Geoffrey Kay, *Development and Underdevelopment: A Marxist Analysis* (New York, 1975), pp. 98–105.

CHAPTER II

1. Evaristo San Cristóval, *Manuel Pardo y Lavalle, su vida y su obra* (Lima, 1945), p. 353.
2. Foreign domination of Tarapacá was so complete that the province was known as "the forgotten south" of Peru; see Romero, *Historia económica*, 2, p. 149.
3. Bonilla, *Guano y burguesía*, pp. 26–35, 39–49.
4. Greenhill and Miller, "Nitrate Trade," pp. 108–09; Levin *Export Economies*, pp. 115, 118; Watt Stewart, *Henry Meiggs, Yankee Pizarro* (Durham, N.C., 1946), pp. 298–302.
5. On the destabilizing effects of this phenomenon, see Merle Kling, "Toward a Theory of Power and Political Instability in Latin America," reprinted from *Western Political Science Quarterly* 9 (March 1956),pp. 21–35, in *Latin America: Reform or Revolution?*, ed. James Petras and Maurice Zeitlin (New York, 1968), pp. 76–93.
6. Jorge Basarde, *Historia de la república del Perú*, 5th ed., rev. and enl., 6 vols. (Lima, 1961–62), vol. 5, p. 2039; Stewart, *Meiggs*, p. 306.
7. Aldunate, *Leyes i documentos*, part I, pp. 33–34.

8. St. John to Derby, Lima, 10 June 1875. F.O. 177/148. St. John was the British minister in Lima, and Derby the British foreign secretary.
9. AGS to WGC, London, 31 December 1874. GMS, 11,471/2; Read to WGC at Lima, Iquique, 24 May, 14 July 1875. GMS, 11,123.
10. St. John to Derby, Lima, 10 June 1875. F.O. 177/148.
11. St. John to Derby, Lima, 13 September 1875. F.O. 61/289; Stewart, *Meiggs*, p. 313. Ten million soles of the loan had already been advanced at the time of the agreement.
12. St. John to Derby, Lima, 4 March 1876. F.O. 61/294; Saul, *British Overseas Trade*, pp. 97–99.
13. St. John to Derby, Lima, 12 February, 22 May 1876. F.O. 61/294.
14. Bermúdez, *Historia del salitre*, p. 333; Basarde, *Historia del Perú*, 5, p. 2039.
15. Aldunate, *Leyes i documentos*, part I, p. 37.
16. Bermúdez, *Historia del salitre*, p. 337.
17. Ibid., pp. 333–35.
18. Luis Esteves, *Apuntes para la historia económica del Perú* (Lima, 1882), no p. no. cited, cited by Basarde, *Historia del Perú*, 5, p. 2234.
19. WGC to Manager of T.N.C., Lima, 4 August 1875. GMS, 11,132; WGC at Lima to WGC at Valparaíso, Lima, 7 April 1876. GMS, 11,121.
20. *La Patria* (Valparaíso), 27 May 1875.
21. WGC at Lima to WGC at Valparaíso, Lima, 5 May 1875. GMS, 11,121.
22. Greenhill and Miller, "Nitrate Trade," pp. 118–21.
23. Bohl to WGC at Valparaíso, Lima, 30 May 1876. GMS, 11,121.
24. Read to WGC at Lima, Iquique, 21 April 1875. GMS, 11,123.
25. WGC at Lima to WGC at Valparaíso, Lima, 5 May 1875. GMS, 11,121.
26. Cruchaga, *Salitre y guano*, pp. 280–81; Bohl to WGC at Valparaíso, Lima, 10 December 1875. GMS, 11,121.
27. Read to WGC at Lima, Iquique, 22 June 1874. GMS, 11,123; Annual Accounts of the Tarapacá Nitrate Company, Iquique, 30 April 1875, f. 34. GMS, 11,049 A; also see p. 23 above.
28. AGS to WGC, London, 31 December 1875. GMS, 11,471/2.
29. WGC at Lima to WGC at Valparaíso, Lima, 7 April, 17 May 1876. GMS, 11,121.
30. The real market value of the T.N.C. was approximately 983,000 soles. This estimate is based on data contained in Billinghurst, *Capitales salitreros*, pp. 27–29; Hayne to Henry, Valparaíso, 11 April 1874. GMS, 11,120; Yearly Accounts of the Tarapacá Nitrate Company, Iquique, 31 December 1881, f. 64. GMS, 11,049 A; Yearly Accounts, Iquique, 31 December 1882, 1883, 1884, f. 17. GMS, 11,033 A/1/2.
31. AGS to WGC London, 15 September 1876. GMS, 11,471/8.
32. WGC at Lima to WGC at Valparaíso, Lima, 17 May 1876. GMS, 11,121; Hayne to Henry, Valparaíso, 22 August 1876. GMS, 11,120.
33. Greenhill and Miller, "Nitrate Trade," pp. 120–21; Bohl to WGC at Valparaíso, Lima, 30 May 1876. GMS, 11,121.

34. Miller to Hayne, Valparaíso, 13 July 1877. GMS, 11,470/1.
35. Comber to Hayne, Valparaíso, 6 May, 14 May 1878. GMS, 11,120.
36. Read to Bohl, Iquique, 28 August 1876. GMS, 11,123.
37. Basarde, *Historia del Perú*, 5, p. 2039.
38. WGC at Lima to WGC at Valparaíso, Lima, 6 May 1876. GMS, 11,121.
39. Bohl to WGC at Valparaíso, Lima, 2 June 1876. GMS, 11,121; Hernández, *Salitre*, p. 89.
40. U.S. Department of State, *Papers Relating to the Foreign Relations of the United States Transmitted with the Annual Message of the President to Congress* (Washington, D.C., 1876), Gibbs to Fish, Lima, 13 July 1876, pp. 420–22. Gibbs was the U.S. minister to Peru, and Fish was the U.S. secretary of state. Hayne to Gibbs, Lima, 27 July 1876. GMS, 11,470/1.
41. Hayne to Gibbs, Lima, 27 July 1876. GMS, 11,470/1.
42. *NV*, vol. 195, 5 January 1876. f. 21.
43. Compañía Salitrera Valparaíso, *Memoria* (Valparaíso, 1877), pp. 3–4, 6.
44. On America, see *NV*, vol. 192, 21 July 1875, fs. 42–43; *NV*, vol. 195, 10 January 1876, fs. 37–38; *NV*, vol. 196, 30 December 1876, fs. 934–36; *NV*, vol. 202, 15 October 1877, fs. 494–504. On California, see *NV*, vol. 191, 15 June 1875, fs. 439–40; *JI*, legajo 387, pieza 8, 1876, fs. 1–80; *NV*, vol. 204, 16 November 1877, fs. 487–96; *NV*, vol. 216, 2 May 1879, fs. 363–65. On Solferino, see *JI*, legajo 74, pieza 1, 1876, fs. 1–153.
45. *MRE*, vol. 176, Antonio Solari Millas to Señor Ministro de Relaciones Esteriores, Iquique, 15 May 1876. Solari was the Chilean consul in Iquique.
46. F. Rondizzoni, *Minerales, guano i salitre de Atacama. Medidas oficiales para el fomento de la industria* (Santiago, 1877), pp. 44–46.
47. Harold Blakemore, *British Nitrates and Chilean Politics, 1886–1896: Balmaceda & North* (London, 1974), pp. 46–47.
48. Derby to Graham, Foreign Office, 16 February, 7 April, 23 April 1877. Graham to Derby, Lima, 27 February, 13 March 1877. F.O. 61/294. Graham was the British minister to Peru. St. John to Marquis of Salisbury, Lima, 26 November 1879. F.O. 61/319. Salisbury was the British foreign secretary. AGS to WGC, London, 2 June 1886. GMS, 11,471/24. European control of the railway was not necessarily advantageous to European nitrate producers. Throughout the 1870s *salitreros* of every nationality engaged in bitter disputes with the railway company over questions of freight charges and inefficient service; see, for example, *South Pacific Times* (Callao), 6 August 1872; Smail to Hayne, Oficina Limeña, 3 September 1879. GMS, 11,472/2.
49. Bermúdez, *Historia del salitre*, pp. 338–39. By January 1876, sales agreements had been signed for thirty-six *oficinas de maquina* representing about 61 percent of total productive capacity in the province,, see *El Mercurio* (Valaparaíso), 2 February 1876; Cruchaga, *Salitre y guano*, p. 292; Billinghurst, *Capitales salitreros*, pp. 21-23.
50. Aldunate,*Leyes i documentos,* part I, pp. 41–42.
51. Greenhill and Miller, "Nitrate Trade," p. 125.

52. U.S. Department of State, *Foreign Relations*, Gibbs to Fish, Lima, 13 July 1876, pp. 420–22.
53. Levin, *Export Economies*, p. 108.
54. Stewart, *Meiggs*, pp. 317–26.
55. U.S. Department of State, *Foreign Relations*, Gibbs to Fish, Lima, 13 July 1876, pp. 420–22; St. John to Derby, Lima, 4 March 1876. F.O. 61/294.
56. Hayne to Gibbs, Lima, 7 November 1876. GMS, 11,470/1; Bohl to Comber, Lima, 16 November 1877. GMS, 11,121; Read to Bohl, Iquique, 28 August 1876. GMS, 11,123; *MH*, vol. 1091, table no. 2.
57. AGS to WGC, London, 13 June 1877. GMS, 11,471/4; Read to Bohl, Iquique, 28 August 1876. GMS, 11,123; *MH*, vol. 1091, table no. 2.
58. Bohl to Hayne, Lima, 30 December 1876. GMS, 11,121; Bohl to Read, Lima, 12 October 1878. GMS, 11,132; Stewart, *Meiggs*, p. 332; Basarde, *Historia del Perú*, 5, p. 2234; *MH*, vol. 1091, table no. 2.
59. *NV*, vol. 192, 4 September 1875, fs. 183–85; *NV*, vol. 203, 13 January 1877, fs. 34–42; *NI*, vol. 54, 13 July 1880, fs. 95–96; Smail to Bohl, Iquique, 29 July 1880. GMS, 11,472/4.
60. *NV*, vol. 209, 5 March 1878, fs. 239–40; *NI*, vol. 69, 20 September 1882, fs. 642–45; Hayne to Bohl, Valparaíso, 17 August 1878. GMS, 11,120.
61. *MH*, vol. 1091, table no. 2.
62. *El Mercurio* (Valparaíso), 1 August 1876.
63. *El Mercurio* (Valparaíso), 4 July 1876.
64. Edwards to Hayne, Valparaíso, 1 April 1874. GMS, 11,128.
65. *BSOD*, 1875, pp. 40–41; Drummond Hay to Derby, Valparaíso, 26 September 1876. F.O. 16/189.
66. Bauer, *Rural Society*, p. 69; *Statist*(London), 19 January 1889. For a detailed treatment of the crisis of 1878, see William F. Sater, "Chile and the World Depression of the 1870s," *JLAS* 11 (May 1979), pp. 67–99.
67. AGS to WGC, London, 16 November 1878. GMS, 11,471/7.
68. Hayne to Miller, Lima, 30 October 1878. GMS, 11,121. The lower figure of total contracted production that Hayne cited probably did not include a contract for nitrate to be produced in the Toco district of the Atacama Desert. On the Toco grounds, see Bermúdez, *Historia del salitre*, pp. 359–61; Bohl to WGC, Lima, 2 June 1876. GMS, 11,121. Aldunate, *Leyes i documentos*, part II, pp. 67–72.
69. Bohl to WGC at Valparaíso, Lima, 30 January 1878. GMS, 11,121.
70. Smail to Bohl, Oficina Limeña, 8 April 1878. GMS, 11,123; Bohl to Comber, Lima, 12 December 1877. GMS, 11,121; Smail to WGC at Valparaíso, Oficina Limeña, 31 March 1879. GMS, 11,049 A; Greenhill and Miller, "Nitrate Trade," pp. 126–27.
71. Smail to WGC at Valparaíso, Oficina Limeña, 31 March 1879. GMS, 11,049 A; Smail to Comber, Oficina Limeña, 10 March 1880. GMS, 11,472/3; Bohl to Comber, Lima, 16 November 1877. GMS, 11,121.
72. *MH*, vol. 1091, table no. 2.

73. Annual Accounts of the Tarapacá Nitrate Company, Iquique, 30 April 1877, f. 3, 31 December 1877, f. 3., 31 December 1878, f. 4. GMS, 11,049 A.
74. Hayne to Comber, Lima, 1 May 1878. GMS, 11,121.
75. Greenhill and Miller, "Nitrate Trade," p. 127.
76. Smail to Read, Oficina Limeña, 6 August 1879. GMS, 11,123.
77. Memorandum, "Purchase and Sale of Nitrate Certificates," Lima, 12 July 1879. GMS, 11,129.
78. *JI,* legajo 74, pieza 1, 1876; AGS to W. T. Morrison, London, 16 November 1881. GMS, 11,471/3. Morrison was the manager of the Bank of London, Mexico and South America. Hayne to Bohl, Valparaíso, 3 August 1878. GMS, 11,120; *MH,* vol. 1091.
79. Hernández, *Salitre,* p. 91.
80. Read to Bohl, Oficina Limeña, 6 January 1879. GMS, 11,123.
81. *PP,* 1878, vol. 75, Valparaíso, 4 March 1878, "Report by Consul Drummond Hay on the Trade and Commerce of Valparaíso for the Years 1876 and 1877," p. 1475. Valparaíso, 6 May 1878,"Supplementary Report by Consul Drummond Hay on the Trade and Commerce of Valparaíso for the year 1877," p. 1480.

CHAPTER III

1. Ernest Mandel, *Late Capitalism,* trans. Joris De Bres (London, 1975), pp. 58–59.
2. Brian Loveman, *Chile: The Legacy of Hispanic Capitalism* (New York, 1979), pp. 159–60; Blakemore, *British Nitrates,* pp. 13–14; Leland R. Pederson, *The Mining Industry of the Norte Chico, Chile* (Evanston, Ill. 1966), pp. 187–225; Thomas C. Wright, "Agriculture and Protectionism in Chile, 1880–1930," *JLAS* 7 (May 1975), p. 49.
3. Figueroa, *Diccionario,* III, p. 19.
4. Bauer, *Rural Society,* p. 181.
5. Ibid., p. 185.
6. *Memorias* of joint-stock companies from this period indicate such a crossover pattern, but the limited number of such documents that have survived make it difficult to guage the extent of the pattern.
7. Bauer, *Rural Society,* pp. 91, 120.
8. Benjamín Vicuña Mackenna, *El libro del cobre i del carbón de piedra en Chile* (Santiago, 1883), p. 528.
9. Mamalakis, *Chilean Economy,* pp. 20, 30.
10. On the conflict over church influence in education, see Allen L. Woll, "For God or Country: History Textbooks and the Secularization of Chilean Society, 1840–1870," *JLAS* 7 (May 1975), pp. 23–43. On the problems resulting from rationalization of state functions, see Barrington Moore, Jr., *Social Origins of Dictatorship and Democracy: Lord and Peasant in the Making of the Modern World* (Boston, 1967), p. 439.
11. Valenzuela, *Political Brokers,* p. 189; Cristián A. Zegers, *Aníbal Pinto: Historia política de su gobierno* (Santiago, 1969), passim.

12. Mamalakis, *Chilean Economy*, p. 34.
13. *PP*, 1876, vol. 73, "Report by Mr. Rumbold on the Progress and General Condition of Chile," p. 356.
14. Frank Whitson Fetter, *Monetary Inflation in Chile* (Princeton, 1931), p. 17; *PP*, 1878, vol. 75, "Report by Consul Grierson on the Trade and Commerce of Coquimbo for the year 1877," p. 1464.
15. Bauer, *Rural Society*, pp. 67–73; for an emphatic statement of this argument see Arnold Bauer, review of *The Heroic Image in Chile: Arturo Prat, Secular Saint*, by William F. Sater, in *JLAS* 7 (May 1975), pp. 159–61.
16. *Resúmen de la hacienda pública de Chile desde 1833 hasta 1914* (London 1914), p. 94.
17. For a detailed discussion of the economic crisis, see William F. Sater, "Chile and the World Depression of the 1870s," *JLAS* 5 (May 1979), pp. 67–99.
18. Encina, *Historia de Chile*, XVI, p. 72; Martner, *Política comercial*, 2, pp. 342–54.
19. Bermúdez, *Historia del salitre*, pp. 290–97; Encina, *Historia de Chile*, XVI, p. 110.
20. Martner, *Política comercial*, 2, p. 360; Encina, *Historia de Chile*, XVI, pp. 70–71; *Hacienda pública*, pp. 26–27.
21. *FN*, vol. 413, pieza 14a, Alberto Blest Gana to Aníbal Pinto, Paris, 25 January 1878.
22. *FN*, vol. 413, pieza 14a, Blest Gana to Pinto, Paris, 3 May 1878.
23. Fetter, *Inflation*, p. 27; Encina, *Historia de Chile*, XVI, pp. 75–76.
24. It was not until the outbreak of the War of the Pacific in 1879 that congress approved a watered-down version of this direct tax legislation; see William F. Sater, "Economic Nationalism and Tax Reform in Late Nineteenth Century Chile," *TA* 33 (October 1976), pp. 324–26.
25. *FN*, vol. 413, pieza 14a, Blest Gana to Pinto, Paris, 13 December 1878.
26. Aníbal Pinto Santa Cruz, *Chile, un caso de desarrollo frustrado* 3d ed. (Santiago, 1973), p. 55.
27. Encina, *Historia de Chile*, XV, pp. 134–35.
28. William Jefferson Dennis, *Tacna and Arica: An Account of the Chile-Peru Boundary Dispute and of the Arbitrations of the United States* (New Haven, Conn., 1931; reprint ed., Hamden, Conn., 1967), pp. 1–9, 40; Gonzalo Bulnes, *Guerra del Pacífico*, 3 vols. (Valparaíso, 1911–19), I, pp. 37–39.
29. Robert N. Burr, *By Reason or Force: Chile and the Balancing of Power in South America, 1830–1905* (Berkeley, 1965), pp. 124, 130.
30. Ibid., pp. 134–36.
31. J. Valerie Fifer, *Bolivia: Land, Location and Politics since 1825* (Cambridge, England, 1972), p. 57; Bulnes, *Guerra*, I, pp. 52–53.
32. O'Brien, "Antofagasta Company," pp. 16–17.
33. Cf. Walter La Feber, *The New Empire: An Interpretation of American Expansion, 1860–1899* (Ithaca, N.Y., 1971), passim.
34. Thomas McLeod Bader, "A Willingness to War: A Portrait of the Republic of Chile During the Years Preceding the War of the Pacific," (Ph.D. diss. Univer-

sity of California, Los Angeles, 1967), pp. 194–506.
35. Burr, *Reason or Force,* p. 141; *El Ferrocarril* (Santiago), 5 January 1880; *El Mercurio* (Valparaíso), 17 January 1880.
36. Harold Blakemore, "Limitations of Dependency: An Historian's View and Case Study," *BELC* 18 (June 1975), pp. 78–79.
37. *PP,* 1882, vol. 18, "Correspondence Between Her Majesty's Government and the Governments of Peru and Chile Respecting the Claims of the Peruvian Bondholders" (cited hereafter as "Bondholders"), Drummond Hay to Salisbury, Santiago, 30 August 1879, pp. 14–15; James Fergusson and George Henry Hopkinson to Salisbury, London, 2 December 1879, p. 17; Sir Charles Russell to Lord Tenterden, London, 31 December 1879, p. 19. Fergusson and Hopkinson were trustees for the bondholders, and Russell was the chairman of their committee. Tenterden was a Foreign Office official.
38. *PP,* 1882, vol. 81, "Bondholders," Russell to Tenterden, London, 31 December 1879, p. 19, Fergusson and Hopkinson to Salisbury, London, 2 December 1879, p. 17.
39. *PP,* 1882, vol. 81, "Bondholders," Salisbury to Pakenham, Foreign Office, 12 December 1879, p. 18.
40. *PP,* 1882, vol. 81, "Bondholders," Russell to Tenterden, London, 31 December 1879, p. 19, Salisbury to Blest Gana, Foreign Office, 5 January 1880, p. 19.
41. *PP,* 1882, vol. 81, "Bondholders," "Declaration made on behalf of the Supreme Government of Chile... to the Peruvian Bondholders' Committee,..." Copy encl. in Blest Gana to Salisbury, Paris, 31 January 1880, pp. 20–21.
42. *PP,* 1882, vol. 81, "Bondholders," Pakenham to Luis Amunátegui, Valparaíso, 29 January 1880, p. 23.
43. Russell to Salisbury, London, 23 January 1880. F.O. 61/323.
44. Russell to Salisbury, London, 3 February 1880. F.O. 61/323.
45. *PP,* 1882, vol. 81, "Bondholders," Amunátegui to Pakenham, Valparaíso, 8 February 1880, pp. 23–24.
46. Augusto Matte to Rafael Sotomayor, n.p., 6 February 1880, quoted in Bulnes, *Guerra,* II, p. 60. Sotomayor was the Chilean minister of war.
47. Pakenham to Salisbury, Valparaíso, 17 January 1880. F.O. 16/207.
48. "Contract between the Supreme Government of Peru and the Société Générale de Crédit Industriel et Commercial...," London, January 1880. Copy annexed to memorandum of J. Pauncforte, Foreign Office 23 January 1880. F.O. 61/323.
49. Herbert Millington, *American Diplomacy and the War of the Pacific* (New York, 1948), p. 98; Dennis, *Tacna and Arica,* p. 146; Frederick B. Pike, *Chile and the United States, 1880–1962: The Emergence of Chile's Social Crisis and the Challenge to United States Diplomacy* (Notre Dame, Ind., 1963), p. 49.
50. Pike, *Chile and the United States,* p. 49.
51. Hernández, *Salitre,* p. 100; Oscar Bermúdez Miral, "El salitre de Tarapacá y Antofagasta durante la ocupación militar chilena," *Anales de la Universidad del Norte* (Antofagasta) 5 (1966), pp. 132–33, 137.

52. Smail to Comber, Iquique, 18 December 1879. GMS, 11,472/3; Bermúdez, "Salitre de Tarapacá," p. 139.
53. Smail to Comber, Oficina Limeña, 28 February 1880. Smail to Comber, Iquique, 8 March 1880 GMS, 11,472/3; Bermúdez, "Salitre de Tarapacá," p. 140.
54. Gibbs and Company (cited hereafter as GC) to Pakenham, Valparaíso, 2 March 1880; Pakenham to Amunátegui, Valparaíso, 3 March 1880; Amunátegui to Pakenham, Santiago, 8 March 1880; Pakenham to Amunátegui, Valparaíso, 8 March 1880; copies encl. in Pakenham to Salisbury, Valparaíso, 24 March 1880. F.O. 16/207.
55. Pakenham to Amunátegui, Valparaíso, 19 March 1880; copy encl. in Pakenham to Salisbury, Valparaíso, 24 March 1880. F.O. 16/207.
56. Amunátegui to Pakenham, Santiago, 23 March 1880; copy encl. in Pakenham to Salisbury, Valparaíso, 26 April 1880. F.O. 16/207.
57. GC to Pakenham, Valparaíso, 30 March 1880; copy encl. in Pakenham to Salisbury, Valparaíso, 26 April 1880. F.O. 16/207. Gibbs's underscoring.
58. Amunátegui to Pakenham, Santiago, 4 May 1880; GC to Pakenham, Valparaíso, 12 May 1880; Copies encl. in Pakenham to Salisbury, Valparaíso, 22 May 1880. F.O. 16/207.
59. Billinghurst, *Capitales salitreros,* p. 39.
60. AGS to GC, London, 20 May 1880. GMS, 11,471/9.
61. GC to AGS, Valparaíso, 10 February 1881, 24 February 1881. GMS, 11,470/4.
62. Smail to Bohl, Oficina Limeña, 6 July 1880. Smail to Comber, Oficina Limeña, 12 November 1880. GMS, 11,472/4.
63. Thomas Morrison to George Rose Innes, London, 19 November 1880. GMS, 11,471/10. Innes was a British merchant who had long been involved in Chilean trade.
64. George Gibbs to Comber, London, 16 March 1880. GMS, 11,471/9. George Gibbs was one of the senior partners of Anthony Gibbs and Sons.
65. Aldunate, *Leyes i documentos,* part I, pp. 179–81; Hernández, *Salitre,* pp. 102–3, 114, 116.
66. Aldunate, *Leyes i documentos,* part I, pp. 83–87.
67. *BSOS,* 1888, pp. 226, 243.
68. Aldunate, *Leyes i documentos,* part I, p. 75.
69. St. John to Phillip W. Currie, London, 10 November 1882. F.O. 61/357. Currie was an assistant undersecretary of state.
70. "Las salitreras de Taltal," *Boletín de la Sociedad Nacional de Agricultura* (Santiago), n.d., reprinted in *El Mercurio* (Valparaíso), 18 August 1879.
71. *PP,* 1880, vol. 74, "Report by Vice-Consul Mark on the Trade and Commerce of Caldera for the year 1879," p. 1522.
72. *BSOS,* 1879, p. 147; *La Patria* (Valparaíso), 9 July 1879; GC to AGS, Valparaíso, 3 March 1884. GMS, 11,470/7.
73. O'Brien, "Antofagasta Company," p. 20.
74. Ibid., pp. 19–20.
75. *El Mercurio* (Valparaíso), 6 February 1880.

76. *PP*, 1881, vol. 91, "Report by Consul Drummond Hay on the Trade and Commerce of Valparaíso for the Years 1879 and 1880," p. 1453.
77. Francisco Vidal Gormaz, *Estudio sobre el puerto de Iquique* (Santiago, 1880), p. 14; WGC to AGS, Valparaíso, 27 February 1881. GMS, 11,470/5.
78. *La Patria* (Valparaíso), 17 June 1880.
79. Bermúdez, "El salitre de Tarapacá," p. 154.
80. *BSOD*, 1880, p. 472.
81. Ibid., pp. 470–71.
82. Ibid., p. 465.
83. Ibid., p. 383.
84. Ibid., p. 461.
85. See Table 3, above.
86. *BSOD*, 1880, p. 471.
87. *BSOS*, 1880, p. 315.
88. Ibid., pp. 269–71; Hayne to George Gibbs, Valparaíso, 14 February, 28 July, 12 August 1879. GMS, 11,470/3.
89. Hernández, *El salitre*, pp. 97, 112, 115, 117; Semper and Michels, *Industria del salitre*, pp. 136–37; *BSOD*, 1880, p. 625.
90. Serious efforts to remove the export duty on mineral products began in 1878. In 1879 the Chamber of Deputies postponed consideration of the reform until the nitrate question had been resolved; see Vicuña MacKenna, *El libro del cobre*, p. 519; *BSOD*, 1879, pp. 157–58.
91. *La Patria* (Valparaíso), 17 July 1880.

CHAPTER IV

1. Chile, Ministerio de Hacienda, *Memoria del Ministro de Hacienda presentada al congreso nacional de 1882* (cited hereafter as *Memoria de...* (Santiago, 1882), p. 14.
2. Billinghurst, *Capitales salitreros*, p. 49; "Memorandum of the Italian Committee of Peruvian Certificate Holders," (translation from Spanish), n.p., n.d., p. 4. Copy encl. in Arthur V. Marras, Secretary of the Committee of English Certificate Holders to Granville, London, 8 March 1884. F.O. 61/357.
3. Bermúdez, "Salitre de Tarapacá," p. 133.
4. *MH*, vol. 1240, Jefe Político de Tarapacá to Señor Ministerio de Hacienda, Iquique, 26 April 1882.
5. On continued United States interest in the nitrate industry, see Millington, *American Diplomacy*, pp. 137-38.
6. This figure is a projection based on production statistics for seven of the twelve months, see Chile, Ministerio de Hacienda, *Memoria de 1883* (Santiago, 1883), table following p. 41; *La Industria* (Iquique), 11 May 1883; *El Veintiuno de Mayo* (Iquique), 10 July, 4 August, 14 November 1883, 15 January 1884. On state nitrate properties, see *MH*, vol. 1585, Intendente de Tarapacá to Señor Minis-

terio de Hacienda, Iquique, 13 November 1886; Bermúdez, "Salitre de Tarapacá," p. 181.
7. Semper and Michels, *Industria del salitre*, p. 137; Table 1, above.
8. Chile, Ministerio de Hacienda, *Memoria de 1884* (Santiago, 1884), p. 1xviii.
9. Bermúdez, "Salitre de Tarapacá," p. 178.
10. See Table 3, above.
11. GC to AGS, Valparaíso, 11 February 1882. GMS, 11,470/5.
12. This figure is a projection based on production statistics for seven of the twelve months; see Chile, Ministerio de Hacienda, *Memoria de 1883*, table following p. 41; *La Industria* (Iquique), 11 May 1883; *El Veintiuno de Mayo* (Iquique), 10 July, 4 August, 14 November 1883, 15 January 1884.
13. GC to AGS, Valparaíso, 29 April 1882. GMS, 11,470/5.
14. *NI*, vol. 62, 25 July 1881, fs. 237–38; *NI*, vol. 69, 20 September 1882, fs. 642–45, 28 November 1882, fs. 892–95, 30 November 1882, fs. 906–9.
15. GC to AGS, Valparaíso, 11 February 1882. GMS, 11,470/5.
16. Smail to GC at Valparaíso, Oficina Limeña, 19 July 1880. GMS, 11,472/4; Chile, Ministerio de Hacienda, *Memoria de 1883*, table following p. 41.
17. "Memorandum... Peruvian Certificate Holders," p. 2. Copy encl. in Marras to Granville, London, 8 March 1884. F.O. 61/357.
18. Ibid., pp. 2–4.
19. While a number of studies of the nitrate industry deal with North and Harvey, by far the best treatment of their early careers is contained in Blakemore, *British Nitrates*, pp. 22–31. Much of the information that follows is based on that work.
20. On the profitability of North's water company, see GC to AGS, Valparaíso, 14 January 1882. GMS, 11,470/5.
21. *NI*, vol. 54, 13 July 1880, fs. 95–96; Smail to Bohl, Iquique, 29 July 1880. GMS, 11,472/4.
22. GC to AGS, Valparaíso, 27 September 1881. GMS, 11,470/4.
23. North himself never did have a very thorough knowledge of the nitrate region. In 1888 one of the Gibbs partners had to use a map of Tarapacá to show North where one of his own properties was located; see AGS to GC, London, 7 September 1888. GMS, 11,471/30.
24. Blakemore, *British Nitrates*, pp. 26–27; Billinghurst, *Capitales salitreros*, pp. 43–44.
25. Billinghurst, *Capitales salitreros*, pp. 48–49.
26. Blakemore, *British Nitrates*, pp. 28–29.
27. Billinghurst, *Capitales salitreros*, p. 28.
28. *NI*, vol. 69, 2 September 1882, fs. 572–76; *NI*, vol. 70, 22 June 1883, fs. 545–48; Blakemore, *British Nitrates*, p. 60.
29. *NI*, vols. 51–71, 1880–83; GC to AGS, Valparaíso, 14 January 1882. GMS, 11,470/5.
30. *NI*, vol. 59, 4 April 1881, fs. 211–13; *NI*, vol. 68, 4 April 1882, fs. 451–52. As of 1882, the Jazpampa was operated by Gibbs under a profit-sharing agreement with North; see Yearly Accounts, Iquique, 31 December 1882, f. 23. GMS, 11,033 A/1.

31. H. B. James to Read, Iquique, 13 May 1879. GMS, 11,472/2.
32. *NI*, vol. 52, 8 June 1880, f. 18.
33. *NI*, vol. 60, 16 August 1881, fs. 507–8.
34. *NI*, vol. 69, 9 December 1882, fs. 929–30.
35. *NI*, vol. 71, 28 December 1883, fs. 1201–3.
36. AGS to GC, London, 30 December 1885. GMS, 11,471/24.
37. Semper and Michels, *Industria del salitre*, pp. 80–81.
38. Chile, Ministerio del Interior, *Memoria del Intendente de Tarapacá correspondiente a 1886* (Santiago, 1887), p. 33. For a detailed discussion of the labor shortage in the 1880s and subsequent decades, see Arthur Lawrence Stickell, "Migration and Mining: Labor in Northern Chile in the Nitrate Era, 1880–1930" (Ph.D. diss., Indiana University, 1979), pp. 47–81.
39. Bauer, *Rural Society*, pp. 163–69; also see Appendix. The low cost of labor in Chile is reflected by the fact that Chilean agricultural laborers earned only about one half as much as their English counterparts; see G. E. Mingay, "The Transformation of Agriculture," in R. M. Hartwell et al., *The Long Debate on Poverty* (London, 1972), p. 44., and Appendix.
40. Bermúdez, *Historia del salitre*, pp. 272–76; Arthur W. Allen, *The Recovery of Nitrate from Chilean Caliche* (Philadelphia, 1921), pp. 1–20.
41. Bermúdez, *Historia del salitre*, p. 276; AGS to GC, London, 3 May 1882. GMS, 11,471/14.
42. Bohl to Hayne, Lima, 16 August 1878. GMS, 11,121; Hayne to Smail, Valparaíso, 29 August 1879. GMS, 11,122.
43. Billinghurst, *Capitales salitreros*, p. 19; Castle, *Iquique*, p. 23.
44. AGS to GC, London, 3 May 1883. GMS, 11,471/14 GC to AGS, Valparaíso, 11 December 1883. GMS, 11,470/7; *El Veintiuno de Mayo* (Iquique), 14 May 1884.
45. T.N.C. Manager's Report, 10 April 1876. GMS, 11,049 A; Yearly Accounts, Iquique, 31 December 1882, f. 34. GMS, 11,033 A/1.
46. GC to AGS, Valparaíso, 14 March 1884. GMS, 11,470/7.
47. Yearly Accounts Iquique, 31 December 1884, f. 45. GMS, 11,033 A/2; GC to AGS, Valparaíso, 9 May 1885. GMS, 11,470/9; Yearly Accounts Iquique, 31 December 1887, f. 42. GMS, 11,033 A/5.
48. GC to AGS, Valparaíso, 24 January 1885. GMS, 11,470/8.
49. Billinghurst, *Capitales salitreros*, p. 61; Smail to Read, Oficina Limeña, 25 June 1880. GMS, 11,472/4.
50. "Estimated Production of Nitrate in the province of Tarapacá with Approximate Cost thereof in Cancha and Placed Alongside of Launches" (cited hereafter as "Production of Nitrate"), table from unnamed source in Liverpool, n.d. copy encl. in AGS to GC, London, 3 May 1882. GMS, 11,471/14; *El Veintiuno de Mayo* (Iquique), 14 May 1884.
51. Castle *Iquique*, p. 42.
52. *NI*, vol. 52, 5 August 1880, fs. 87–92; *NI*, vol. 57, 1 July 1881, f. 5; *NI*, vol. 71, 15 October 1883, fs. 958–60; *NI*, vol. 75, 9 August 1884, fs. 899–901. The International Mercantile Bank had previously been known as the Anglo Peruvian Bank. The name change was no doubt necessitated by the Chilean

occupation of Tarapacá. On the financial problems of the bank, see GC to AGS, Valparaíso, 27 December 1881. GMS, 11,470/5.
53. Billinghurst, *Capitales salitreros,* p. 61; "Production of Nitrate," copy encl. in AGS to GC, London, 3 May 1882. GMS, 11,471/14. *El Veintiuno de Mayo* (Iquique), 14 May 1884; *NI,* vol. 71, 28 December 1883, fs. 1201–3.
54. *NI,* vol. 59, 15 July 1881, fs. 439–43; *NI,* vol. 60, 4 August 1881, fs. 480–84; *NI,* vol. 68, 5 April 1882, fs. 98–99; *NI,* vol. 71, 23 November 1883, fs. 1083–85.
55. GC to AGS, Valparaíso, 27 September 1881. GMS, 11,470/4; GC to AGS, Valparaíso, 11 February 1882. GMS, 11,470/5; Blakemore, *British Nitrates,* p. 32.
56. Blakemore, *British Nitrates,* pp. 32–33.
57. Semper and Michels, *Industria del salitre,* pp. 278–79; *MH,* no vol. no., Delegación Fiscal de Tarapacá, 1880, table no. 2.
58. Yearly Accounts Iquique, 31 December 1883, f. 31. GMS, 11,033 A/1.
59. *MH,* no vol. no., Delegación Fiscal de Tarapacá, 1880, table no. 2.; *MH,* no vol. no., Delegación Fiscal de Tarapacá, 1880–89, table no. 3. The figures for 1878 are actually somewhat low, since they are based only on *oficinas* actually sold to the Peruvian government.
60. J. R. Brown, "Nitrate Crises, Combinations and the Chilean Government in the Nitrate Age," *HAHR* 43 (May 1963), p. 232.
61. *MH,* vol. 1089, Eduardo Délano to J. M. Carreño, Gobernado Civil de Tarapacá, Iquique, 18 February 1880. Robert Harvey to Carreño, Iquique, 18 February 1880; *NI,* vol. 69, 7 September 1882, fs. 597–610; *NI,* vol. 71, 4 October 1883, fs. 916–23; on the *oficina's* high production costs, see "Production of Nitrate," copy encl. in AGS to GC, London, 3 May 1882. GMS, 11,471/14.
62. *NI,* vol. 69, 20 September 1882, fs. 642–45; *NI,* vol. 71, 4 October 1883, fs. 916–23.
63. GC to AGS, Valparaíso, 27 December 1881. GMS, 11,470/5; GC to AGS, Valparaíso, 11 December 1883. GMS, 11,470/7.
64. *NI,* vol. 54, 13 August 1880, fs. 133–34; *NI,* vol. 60, 29 July 1881, fs. 462–65; *NI,* vol. 76, 14 October 1884, document no. 85 A; *NI,* vol. 71, 14 November 1883, fs. 1051–53; *NI,* vol. 80, 15 October 1885, fs. 1301–03; "Production of Nitrate," copy encl. in AGS to GC, London, 3 May 1882. GMS, 11,471/14.

CHAPTER V

1. *BSES,* 1881–1882, *p. 178.*
2. Wright, "Agriculture and Protectionism," p. 47; Bauer, *Rural Society* p. 71.
3. Figures on the labor force are averages based on monthly data in Chile, Ministerio de Hacienda, *Memoria de 1885* (Santiago, 1885), p. 5; Chile, Ministerio de Hacienda, *Memoria de 1890–1891* (Santiago, 1891), p. 251. On the importation of agricultural products, see Smail to Bohl, Oficina Limeña, 13 June 1878. GMS, 11,472/1; Billinghurst, *Capitales salitreros,* p. 110.
4. Wright, "Agriculture and Protectionism," p. 49.

5. *PP*, 1888, vol. 100, "Report on the Trade and Commerce of Chile for the Year 1886," p. 3.
6. *PP*, 1888, vol. 100, "Agriculture in Chile," p. 2. For an earlier reference to the significance of the northern market, see *BSES*, 1883–84, pp. 97, 114.
7. Wright, "Agriculture and Protectionism," p. 49.
8. Subercaseaux, *Banking Policy*, pp. 202–3.
9. Ibid., p. 97.
10. Bauer, *Rural Society*, pp. 101–4, 152; Thomas C. Wright, "The Politics of Inflation in Chile, 1888–1918," *HAHR* 53 (May 1973), p. 243.
11. Bauer, *Rural Society*, pp. 159–61.
12. Wright, "Agriculture and Protectionism," p. 58.
13. Moore, *Dictatorship and Democracy*, p. 434.
14. Valenzuela, *Political Brokers*, pp. 190–91.
15. Bauer, *Rural Society*, pp. 215–17.
16. Ibid., pp. 90–96.
17. Hernández, *Salitre*, p. 177.
18. *Hacienda publica*, pp. 48–49.
19. Santiago Marín Vicuña, *Los ferrocarriles de Chile*, 3 ed. (Santiago, 1912), pp. 138–39; Martner, *Política comercial*, II, p. 449.
20. For the best summary of Balmaceda's economic policies, see Hernán Ramírez Necochea, *Balmaceda y la contrarrevolución de 1891*, 3d ed. (Santiago, 1972), pp. 113–63. For additional details, see Blakemore, *British Nitrates*, pp. 69–75.
21. Martner, *Política comercial*, II, p. 449; Ramírez, *Balmaceda*, pp. 125–26.
22. Mamalakis, *Chilean Economy*, p. 75.
23. Encina, *Historia de Chile*, XVIII, p. 384; Blakemore, *British Nitrates*, p. 72.
24. Valenzuela, *Political Brokers*, p. 195; Blakemore, *British Nitrates*, p. 174; Marín Vicuña, *Ferrocarriles*, p. 91; Martner, *Política comercial*, II, p. 449. Statistics on public employment include only white-collar workers as evidenced by the fact that in 1884 the state railways alone employed 5,300 workers; see Marín Vicuña, *Ferrocarriles*, p. 97.
25. Mamalakis, *Chilean Economy*, p. 77; *La Epoca* (Santiago), 6 June 1884; *PP*, 1890, vol. 73, "Report on European Emigration to Chile," p. 7.
26. Semper and Michels, *Industria del salitre*, pp. 157–58.
27. Santelices, *Bancos*, p. 263.
28. *El Veintiuno de Mayo* (Iquique), 15 December 1887; *NI*, vols. 53–118, 1880–90; GC to AGS, Valparaíso, 9 January 1885. GMS, 11,470/8.
29. *MH*, vol. 1043, Gonzalo Bulnes, Jefe Político de Tarapacá to Señor Ministro de Hacienda, Iquique, 4 September 1884; GC to AGS, Valparaíso, 2 October 1883. GMS, 11,470/7.
30. *MH*, vol. 1043, P. N. Gandarillas to Señor Ministro de Hacienda, n.p., 6 August 1884. Gandarillas was the director of the treasury. The arrangement with the bank was altered by a decree of 7 July 1884 that allowed *salitreros* to deposit duty payments in the form of bank drafts in the bank's Valparaíso

branch. This volume of the Archivo del Ministerio de Hacienda contains protests by the bank over the new decree.
31. *El Veintiuno de Mayo* (Iquique), 11 December 1887.
32. Billinghurst, *Capitales salitreros,* p. 63; *ElVeintuino de Mayo* (Iquique), 15 December 1887.
33. Salvador Soto Rojas, *Las riquezas de Chile en sus industrias y comercio* (Santiago, 1906), pp. 18–19; Santelices, *Bancos,* pp. 183, 222, 248, 268, 274–75; *La Patria* (Valparaíso), 27 January 1875, 13 January 1888.
34. Santelices, *Bancos,* p. 386; Subercaseaux, *Banking Policy,* pp. 98, 190–92.
35. Kirsch, *Industrial Development,* p. 59.
36. On the Banco Mobilario, see Banco Mobilario, *Memoria* (Valparaíso, 1877); on the Banco Nacional, see *La Patria* (Valparaíso), 11 January 1884. Landowners also constituted a significant portion of the shareholders in the Banco de Valparaíso, although their domination does not appear to have been as thorough as in the case of the Banco Nacional; see *MH,* vol. 1364, Bancos Correspondencia, 1883.
37. *PP,* 1887, vol. 83, "Report on the Trade of Valparaíso, 1884–85," p. 7.
38. *Hacienda pública,* p. 92; Chile, Ministerio de Hacienda, *Memoria de 1890* (Santiago, 1890), pp. 199, 207.
39. *La Epoca* (Santiago), 2 December 1884.
40. Kirsch, *Industrial Development,* pp. 5–7, 14–15, 24.
41. Ibid., pp. 7–10, 15.
42. Ibid., pp. 58–95; Wright, "Agriculture and Protectionism," p. 58.
43. *SAJ,* 18 January 1885; GC to AGS, Valparaíso, 15 September 1882. GMS, 11,470/6; *PP,* 1887, vol. 83, "Report on the Trade and Commerce of the Province of Coquimbo (Chile) for the year 1886," p. 1; Martner, *Política comercial,* II, p. 464; *FN* (London), 22 April 1891; Herrmann, *Oro, plata i cobre,* pp. 35–36, 51; Escobar, *Mercado,* p. 52.
44. Henry Sewell Gana, *British Capital and Chilian Industry, Nitrates, Gold Mines and Coal Mines* (London, 1889), pp. 27–31, Escobar, *Mercado,* p. 52.
45. Escobar, *Mercado,* p. 51.
46. Ibid., p. 52.
47. Ibid., p. 52. On the close relationship between stock investments and interest rates, see GC to AGS, Valparaíso, 16 December 1882. GMS, 11,470/6.
48. *SAJ,* 1 October 1887.
49. Information on the stockholders of public companies is available from a variety of sources including the newspapers *La Patria* and *El Mercurio,* which published the semiannual *memorias* of the major banks and insurance companies, as well as occasional *memorias* of other firms. The Biblioteca Nacional has a number of *memorias* from the period, although there are few consecutive runs for any single corporation.
50. GC to AGS, Valparaíso, 11 December 1883. GMS, 11,470/7.
51. *FN* (London), 11 February 1889.
52. GC to AGS, Valparaíso, 3 March 1884. GMS, 11,470/7.

53. Ibid.
54. Wright, "Agriculture and Protectionism," p. 48.
55. Chile, Ministerio de Hacienda, *Memoria de 1885*, p. 5.
56. Nathan Miers Cox, "Imprevisión de los salitreros," *Boletín de la Sociedad Nacional de Agricultura* (Santiago), n.d., reprinted in *El Mercurio* (Valparaíso), 10 January 1885.
57. *FN*, vol. 414, pieza 2a, Santa María to Blest Gana, Santiago, 2 August 1884.
58. *MH*, vol. 1392, Bulnes to Señor Ministro de Hacienda, Iquique, 12 September 1884.
59. This was a practice instituted in Iquique by the Peruvian government. Since the workers' wages depended on the amount of nitrate loaded, the account books of their union could be relied on as an accurate estimate of the amount of nitrate shipped.
60. Chile, Ministerio del Interior, *Memoria del Intendente de Tarapacá correspondiente a 1886*, p. 55.
61. *Hacienda pública*, p. 92.
62. Bauer, *Rural Society*, pp. 152–53; Chile, Ministerio de Hacienda, *Memoria de 1887* (Santiago, 1887), p. xxv.
63. *MH*, vol. 1585, Intendencia de Tarapacá. This volume contains a letter from *salitreros* in Tarapacá requesting exemption from a law passed in 1852 prohibiting the use of *fichas*.
64. Castle, *Iquique*, p. 12.
65. AGS to GC, London, 16 September 1890, no. 63. GMS, 11,471/36.
66. For a discussion of such symbiotic relationships defined as articulations of modes of production, see Aidan Foster-Carter, "The Modes of Production Controversy," *NLR* 107 (January–February 1978), pp. 47–77.

CHAPTER VI

1. Karl Marx, "The Eighteenth Brumaire of Louis Bonaparte," in Robert C. Tucker, ed., *The Marx-Engels Reader* (New York, 1972), p. 437.
2. On factors that define national elite status, see Diana Balmori and Robert Oppenheimer, "Family Clusters: Generational Nucleation in Nineteenth-Century Argentina and Chile," *Comparative Studies in Society and History* 21 (April 1979), pp. 239–53. I am grateful to Professor Oppenheimer for information on the membership in the Club de la Unión of some of the individuals discussed in this chapter.
3. *NV*, vol. 184, 10 April 1874, f. 196; *NV*, vol. 259, 11 April 1885, f. 274, 7 June 1885, fs. 444–45; *NV*, vol. 268, 6 September 1886, f. 176. Costa's other brother Andrés had also been involved in the nitrate industry serving as president of the Sacramento Company.
4. *NV*, vol. 192, 4 September 1875, fs. 183–85; NV, vol. 290, 5 March 1878, fs. 239–40.
5. *NV*, vol. 241, 15 October 1883, f. 265; *NV*, vol. 239, 26 July 1883, f. 100; *NV*,

vol. 262, 3 October 1883, f. 206; *NV*, vol. 247, 3 June 1884, fs. 475-76.
6. Compañía Beneficiadora de Oro de Quillota, *Estatutos* (Valparaíso, 1881), passim; *NV*, vol. 262, 16 December 1885, fs. 611-12; *NV*, vol. 270, 8 May 1886, f. 54.
7. *NI*, vol. 7, 2 July 1872, no f. no.; *NI*, vol. 9, 28 February 1873, no f. no.; *NV*, vol. 190, 30 December 1875, f. 918; *NV*, vol. 195, 5 January 1876, fs. 19-23.
8. *NV*, vol. 238, 5 January 1883, f. 27; *NV*, vol. 249, 5 June 1884. fs. 402-3.
9. *NV*, vol. 237, 3 July 1882, f. 1367.
10. *NV*, vol. 259, 24 October 1885, fs. 386-87; La Protectora Compañía Chilena de Seguros, *Estatutos* (Valparaíso, 1886), pp. 3, 16.
11. *NV*, vol. 269, 8 January-3 April 1886, fs. 24-556; *NV*, vol. 270, 19 July 1886, fs. 445-46.
12. *NI*, vol. 87, 20 October 1886, fs. 1781-82; Fábrica Nacional de Cerveza, *1a Memoria* (Valparaíso, 1890), pp. 1, 7; Kirsch, *Industrial Development*, p. 39.
13. *NV*, vol. 252, 30 September 1884, fs. 573-74; *NI*, vol. 78, 7 January 1885, fs. 13-15; *NI*, vol. 79, 11 May 1885, fs. 624-26; *NI*, vol. 79, 18 July 1885, fs. 903-4; Castle, *Iquique*, pp. 38-40; GC to AGS, Valparaíso, 5 July 1885. GMS, 11,470/9.
14. Compañía de Gas de Concepción, *Estatutos* (Valparaíso, 1887), p. 4; *NV*, vol. 238, 27 June 1883, document no. 24; *NV*, vol. 240, 26 January 1883, f. 520; 20 March 1883, f. 555, 11 April 1883, f. 563; 18 April 1883, f. 564; Ferrocarril Urbano de Santiago, *16a Memoria* (Santiago, 1882), pp. 33-34.
15. *NV*, vol. 240, 20 March 1883, f. 555, 11 April 1883, f. 563, 18 April 1883, f. 564.
16. *NV*, vol. 236, 25 January 1882, fs. 153-54; *NV*, vol. 270, 13 May 1886, fs. 73-74; *NS*, vol. 634, 29 September 1881, fs. 98-99.
17. *NV*, vol. 236, 25 January 1882, fs. 153-54.
18. On Necochea's loan, see *NV*, vol. 236, 18 May 1882, fs. 589-90; Herrera was linked to the Edwards family in a number of enterprises, including the beer and insurance companies mentioned above, and was a regular customer at the Edwards bank; see, for example, *NV*, vol. 237, 3 July 1882, f. 1367.
19. Figueroa, *Diccionario*, III, p. 18, V, p. 717; *NV*, vol. 269, 16 April 1886, fs. 631-32.
20. Compañía Comercial de Remolcadores, *Estatutos* (Valparaíso, 1881), p. 4; Ferrocarril Urbano de Santiago, *16a Memoria* (Santiago, 1882), p. 3; *NV*, vol. 236, 29 July 1882, fs. 171-84.
21. Compañía de Maderas y Carbón *Estatutos* (Valparaíso, 1883), p. 12; Compañía Telegrafo Transandino, *Estatutos* (Valparaíso, 1887), p. 14; Compañía de Gas de Concepción, *Estatutos*, p. 4; *NV*, vol. 263, 27 May 1886, fs. 751-53.
22. *NV*, vol. 264, 28 December 1886, fs. 943-44; *NV*, vol. 236, 12 April 1882, fs. 100-106; *NV*, vol. 246, 18 November 1884, fs. 651-52; *NV*, vol. 269, 16 April 1886, fs. 631-32.
23. *NV*, vol. 264, 28 December 1886, fs. 943-44; Enrique Espinoza, *Jeografía descriptiva de la república de Chile*, 4th ed. (Santiago, 1897), p. 196.
24. Figueroa, *Diccionario*, V, p. 717.
25. *NI*, vol. 69, 7 September 1882, fs. 597-610; GC to AGS, Valparaíso, 6 August

1886. GMS, 11,470/9; Miller to Herbert Gibbs, Valparaíso, 14 October 1887. GMS, 11,470/10; *NS*, vol. 723, 6 June 1887, fs. 581–82; *NS*, vol. 761, 11 October 1888, fs. 795–98.
26. *NV*, vol. 235, 10 July 1882, f. 533; *NV*, vol. 244, 22 August 1883, f. 1100; *MH* vol. 1364, "Memoria del Banco de Valparaíso," Valparaíso, 13 January 1883.
27. Compañía Refinería de Azucar de Viña del Mar, *Estatutos* (Valparaíso, 1887), p. 21; Fábrica Nacional de Cerveza, *la Memoria*, p. 7; Kirsch, *Industrial Development, p. 184.*
28. *NV*, vol. 797, 25 November 1890, f. 658; Espinoza, *Jeografía*, pp. 229, 352.
29. Figueroa, *Diccionario*, II, p. 553.
30. Figueroa, *Diccionario*, III, pp. 19–20.
31. O'Brien, "Antofagasta Company," pp. 6–12, 26–27.
32. Bauer, *Rural Society* pp. 194–95.
33. O'Brien, "Antofagasta Company," pp. 12, 16–22, 24–25; Sater, "Economic Nationalism," p. 331; Kirsch, *Industrial Development*, p. 197.
34. Figueroa, *Diccionario*, IV, p. 397; Hernández, *El salitre*, pp. 84–86; GC to AGS, Valparaíso, 27 April 1888. GMS, 11,470/10.
35. *BOI*, 8, 1890, p. 826.
36. Bauer, *Rural Society*, p. 200; Figueroa, *Diccionario*, IV, p. 397.
37. Banco Mobilario, *Memoria* (Santiago, 1877), p. 4; *NS*, vol. 625, 5 January 1881, f. 5, 7 January 1881, fs. 11–12, 21 April 1881, f. 164; *NS*, vol. 647, 30 November 1882, fs. 413–14; *NS*, vol. 661, 3 January 1883, fs. 26–27.
38. *NS*, vol. 716, 18 March 1887, fs. 190–92.
39. *NS*, vol. 675, 26 September 1884, f. 793; *NS*, vol. 634, 13 October 1881, fs. 153–54; *NS*, vol. 661, 31 July 1883, fs. 457–60; *NS*, vol. 666, 25 October 1884, fs. 428–32; *NS*, vol. 693, 30 July 1885, fs. 601–2; *NS*, vol. 659, 8 May 1883, fs. 257–59; *NS*, vol. 674, 13 September 1884, fs. 753–54; *NS*, vol. 716, 21 January 1887, fs. 86–87; *NS*, vol. 738, 21 April 1880, fs. 507–8; *NS*, vol. 794, 24 November 1890, fs. 454–59.
40. Figueroa, *Diccionario*, V, p. 869; *NS*, vol. 675, 26 September 1884, f. 793. Records of Subercaseaux's business transactions cited in notes 37–39, above, list Concha y Toro as his *apoderado*, or attorney.
41. Lía Cortés and Jordi Fuentes, *Diccionario político de Chile, 1810–1966* (Santiago, 1967), p. 81; Encina, *Historia de Chile*, XIV, p. 499.
42. Compañía Huanchaca de Bolivia, *7a Memoria* (Valparaíso, 1874), p. 3; *18a Memoria* (Valparaíso, 1890), p. 3; Marín Vicuña, *Ferrocarriles*, p. 39.
43. Kirsh, *Industrial Development*, p. 189.
44. *NS*, vol. 688, 26 March 1885, fs. 349–50; Espinoza, *Jeografía descriptiva*, p. 237.
45. Figueroa, *Diccionario*, II, p. 421; O'Brien, "Antofagasta Company," pp. 17, 19–20, 24.
46. Brian Loveman, *Hispanic Capitalism*, p. 273.
47. Concerning the preconditions for the emergence of mature capitalism, see Ben Fine, "On the Origins of Capitalist Development," *NLR*, 109 (May–June 1978), pp. 88–95.

CHAPTER VII

1. Ramírez, *Balmaceda*, p. 99.
2. Brown, "Nitrate Combinations," pp. 233–34.
3. Karl Marx, *Capital: A Critique of Political Economy*, ed. Frederick Engels, trans. Ernest Untermann (Chicago, 1909), I, p. 688.
4. A. K. Cairncross, *Home and Foreign Investment, 1870–1913* (Cambridge, England, 1953), p. 84.
5. *BOI*, II, 1893, p. 1729.
6. Cairncross, *Investment*, p. 85.
7. Ibid., p. 228.
8. J. Fred Rippy, "The British Investment 'Boom' of the 1880's in Latin America," *HAHR* 29 (May 1949), p. 281. These figures do not include insurance companies, private partnerships, or transoceanic steamship lines.
9. Mandel, *Late Capitalism*, pp. 57–58.
10. Irving Stone, "British Long-Term Investment in Latin America, 1865–1913," *Business History Review* 42 (Autumn 1968), pp. 315–23; Rippy, "Investment 'Boom'," p. 281.
11. *BOI*, 6, 1888, pp. 653, 695.
12. Blakemore, *British Nitrates*, 38–42.
13. J. Fred Rippy, "Economic Enterprises of the 'Nitrate King' and His Associates in Chile," *PHR* 17 (November 1948), pp. 460–61; *BOI*, 7, 1889, p. 810, 8, 1890, p. 866.
14. Billinghurst, *Capitales salitreros*, pp. 49–50.
15. *NI*, vol. 78, 9 January 1885, fs. 17–20; *NI*, vol. 100, 19 May 1888, fs. 749–52; *BOI*, 8, 1890, pp. 866–67.
16. *SAJ*, 12 October 1888.
17. *NI*, vol. 71, 28 December 1883, fs. 1201–3; Billinghurst, *Capitales salitreros*, pp. 82–83.
18. *NI*, vol. 93, 7 June 1887, fs. 806–8; *NI*, vol. 99, 17 April 1888, f. 552; *NI*, vol. 101, 26 December 1888, fs. 1948–50; *BOI*, 8, 1890, p. 888.
19. *BOI*, 8, 1890, p. 882.
20. Miller to Lowe, London, 18 September 1888. GMS, 11,471/31; Miller to Smail, London, 21 January 1889, 9 February 1889. GMS, 11,471/32.
21. Yearly Accounts Iquique, 31 December 1890, fs. 17, 53. GMS, 11,033 A/8; Miller to GC, London, 3 November 1888. GMS, 11,471/31; *BOI*, 8, 1890, p. 907.
22. *FN* (London), 10 January 1889.
23. *SAJ*, 26 January 1889; *FN* (London), 9 February, 1889.
24. Blakemore, *British Nitrates*, pp. 59–60.
25. Ibid., pp. 60–62; David Joslin, *A Century of Banking in Latin America* (London, 1963), pp. 176–81.
26. Blakemore, *British Nitrates* pp. 63–64; Billinghurst, *Capitales salitreros*, pp. 108–11.

27. Blakemore, *British Nitrates,* pp. 47–50; J. R. Brown, "The Chilean Nitrate Railways Controversy," *HAHR* 38 (November 1958), pp. 466–68.
28. GC to AGS, Valparaíso, 20 August 1885, 24 July 1886, 6 August 1886. GMS, 11,470/9; AGS to GC, London, 22 October 1885, 18 November 1885. GMS, 11,471/24.
29. AGS to GC, London, 2 June 1886. GMS, 11,471/24; Blakemore, *British Nitrates,* p. 48.
30. AGS to GC, London, 26 August 1887. GMS, 11,471/27.
31. Blakemore, *British Nitrates,* pp. 48, 50.
32. Marx, *Capital,* III, p. 519.
33. Cairncross, *Investment,* pp. 92–95.
34. Blakemore, *British Nitrates,* pp. 61–62.
35. See, for example, *FN* (London), 24 August 1888.
36. Blakemore, *British Nitrates,* pp. 234–35.
37. Hayne to Smail, London, 20 February 1889. AGS to GC, London, 7 March 1889, no. 5. GMS, 11,471/32; AGS to GC, London, 17 April 1889, no. 8. GMS, 11,471/33.
38. AGS to GC, London, 31 May 1889, no. 13. GMS, 11,471/33.
39. *Statist* (London), 29 November 1890.
40. *FN* (London), 16 October 1889.
41. *BOI,* 6, 1888, 7, 1889, 8, 1890, passim.
42. Paul M. Sweezy, *The Theory of Capitalist Development* (New York, 1968), p. 260.
43. *BOI,* 7, 1889, pp. 550, 704, 760, 762, 796; Rippy, "Nitrate King," p. 460.
44. *SAJ,* 26 July 1890.
45. *SAJ,* 6 April 1889.
46. *FN* (London), 24 October 1889; *SAJ,* 12 October 1889, 1 November 1890; *Economist* (London), 11 January 1890.
47. W. Anderson Smith, *Temperate Chile, A Progressive Spain* (London, 1899), quoted in Joseph Robert Brown, "The Chilean Nitrate Industry in the Nineteenth Century" (Ph.D. diss., Louisiana State University, 1954), pp. 191–92.
48. Chile, Ministerio de Hacienda, *Memoria de 1890,* pp. 38–39.
49. On the subsequent reduction in wages, see AGS to GC, London, 16 September 1890, no. 63. GMS, 11,471/36.
50. *SAJ,* 25 August 1890.
51. *Economist* (London), 11 January 1890.
52. *SAJ,* 14 December 1889, 2 August 1890, 24 October 1891.
53. *SAJ,* 2 November 1889, 5 September 1891; AGS to GC, London, 7 February 1890, no. 40. F. H. Evans to Inglis, London, 24 March 1890. GMS, 11,471/35. Evans was the chairman of the Tamarugal Company.
54. H. L. Gibbs to Vicary Gibbs, London, 10 July 1890. GMS, 11,040/1. Vicary Gibbs was a partner in the London house.
55. *SAJ,* 5 July 1890; *Statist* (London), 12 September 1891.

56. *SAJ*, 4 June 1890,, 8 November 1890, 22 November 1890, 28 November 1891; *FN* (London), 11 November 1890, 17 January 1891.
57. For examples of North's techniques for retaining investor confidence, see Blakemore, *British Nitrates*, p. 42.
58. *SAJ*, 23 November 1889.
59. Blakemore, *British Nitrates*, p. 152.
60. AGS to GC, London, 6 May 1887. GMS, 11,471/26; GC to AGS, Valparaíso, 24 June 1887. GMS, 11,470/10.
61. Daubney to Gibbs, Valparaíso, 5 August 1890. GMS, 11,470/12.
62. On the effect of monopoly capitalism on state apparatus in the capitalist center, see Mandel, *Late Capitalism*, pp. 481–91.

CHAPTER VIII

1. Given the importance of Balmaceda and the Civil War, they have been the subject of innumerable studies. Some of the most significant are: Julio Bañados Espinosa, *Balmaceda: su gobierno y la revolución de 1891*, 2 vols. (Paris, 1894); José Miguel Yrarrázaval Larraín, *El Presidente Balmaceda*, 2 vols. (Santiago, 1940); Encina, *Historia de Chile*, XIX & XX; Ramírez, *Balmaceda*; Blakemore, *British Nitrates*. For an excellent historiographical treatment of the subject, see Harold Blakemore, "The Chilean Revolution of 1891 and Its Historiography," *HAHR* 45 (1965), pp. 393–421.
2. Blakemore, *British Nitrates*, pp. 3–6; Loveman, *Chile*, p. 136.
3. Valenzuela, *Political Brokers*, p. 178.
4. Marcos Kaplan, *Formación del estado nacional en América Latina* (Santiago, 1969), pp. 36–38.
5. Valenzuela, *Political Brokers*, pp. 187–90; on the conflict over church influence in education, see Woll, "For God or Country," pp. 23–43.
6. Blakemore, *British Nitrates*, p. 6.
7. Valenzuela, *Political Brokers*, p. 189. For a more traditional interpretation of the reform process as the triumph of liberalism, see Julio Heise González, *Historia de Chile, el período parlamentario, 1861–1925* (Santiago, 1974), passim.
8. For an interesting example of the transition from liberal reformer to election manipulator involving Senator José Francisco Vergara, see Yrarrázaval, *Balmaceda*, II, pp. 273–80.
9. Ibid., I, pp. 349–50.
10. For the interpretations, see Blakemore, *British Nitrates*, pp. 115–91, and Ramírez, *Balmaceda*, pp. 180–99.
11. Valenzuela, *Political Brokers*, p. 197.
12. Ibid., p. 198.
13. Encina, *Historia de Chile*, XIX, p. 56; Yrarrázaval, *Balmaceda*, I, pp. 318–19.
14. Encina, *Historia de Chile*, XIX, p. 151.
15. *Hacienda pública*, pp. 25–29.

16. Blakemore, *British Nitrates*, p. 75.
17. Yrarrázaval, *Balmaceda*, I, pp. 313-314.
18. Encina, *Historia de Chile*, XIX, p. 41.
19. *Hacienda pública*, pp. 25–27, 92–94; Chile, Ministerio de Hacienda, *Memoria de 1890–1891*, p. 25; Markos J. Mamalakis, "The Role of Government in the Resource transfer and Resource Allocation Processes: The Chilean Nitrate Sector, 1880–1930," in Gustav Ranis, ed., *Government and Economic Development* (New Haven, Conn., 1971), pp. 181–210. Mamalakis's estimate of a one-third return to the private domestic economy is an average for the entire Nitrate Age. Since it assumes that a substantial portion of the nitrate industry was Chilean-owned, the actual return to the domestic economy in the period 1881–90 was no doubt less than one-third of total nitrate revenues.
20. José Miguel Yrarrázaval, *La política económica del presidente Balmaceda* (Santiago, 1963), pp. 77–79.
21. Blest Gana to Pinto, Paris, 30 May 1878, quoted in Alberto Blest Gana, "La situación financiera de Chile en 1878," *Revista Chilena* 13 (April 1922), p. 489.
22. Carlos Walker Martínez, *Historia de la administración Santa María*, 2 vols. (Santiago, 1888–89), I, p. 310.
23. Herrmann, *Oro, plata i cobre*, pp. 48–49.
24. Marín Vicuña, *Ferrocarriles*, pp. 111, 115, 121; *BSOD*, 1889, p. 166.
25. Yrarrázaval, *Balmaceda*, I, p. 315.
26. Walker Martínez, *Santa María*, I, p. 312.
27. Encina, *Historia de Chile*, XIX, pp. 115–17; Yrarrázaval, *Balmaceda*, I, pp. 349, 366.
28. Encina, *Historia de Chile*, XIX, pp. 87–89.
29. Miller to Herbert Gibbs, Steamer Laja near Caldera, 16 January 1888. GMS, 11,470/10.
30. Yrarrázaval, *Política económica*, pp. 74–81.
31. Blakemore, *British Nitrates*, pp. 173–75.
32. Brown, "Nitrate Industry," pp. 58-60, 71; Brown, "Nitrate Railways," p. 467. On the port concession, see p. 71 above.
33. Yrarrázaval, *Política económica*, pp. 15–16; Gonzalo Bulnes, "La combinación salitrera," *La Libertad Electoral* (Santiago), 26 April 1887, reprinted in *El Veintiuno de Mayo* (Iquique), 5 May 1887.
34. Miller to Herbert Gibbs, Valparaíso, 9 January 1888, Steamer Laja near Caldera, 16 January 1888. GMS, 11,470/10.
35. Bulnes, "La combinación salitrera."
36. Miller to Herbert Gibbs, Steamer Laja near Caldera, 16 January 1888. GMS, 11,470/10.
37. *BSOS*, 1888, pp. 35–37, 190–91, 219–46. Criticism of the plan's ineffectiveness was voiced in these sessions, and in an article in *El Mercurio* (Valparaíso), 21 July 1888.
38. Ramírez, *Balmaceda*, p. 97.

39. GC to AGS, Valparaíso, 6 August 1886. GMS, 11,470/9; GC to AGS, Valparaíso, 14 October 1887. GMS, 11,470/10.
40. Charles Jones, "Insurance Companies," in D. C. M. Platt, ed., *Business Imperialism, 1840–1930: An Inquiry Based on British Experience* (Oxford, 1977), p. 57; *El Mercurio* (Valparaíso), 24/29 July 1886; *SAJ*, 7 August 1886; *BSOD*, 1889, pp. 79–82.
41. On the Bank of Tarapacá and London, see the *Statist* (London), 23 March 1889; GC to AGS, Valparaíso, 4 January 1889. GMS, 11,470/11. On the Nitrate Provisions and Supply Company, see Billinghurst, *Capitales salitreros*, pp. 108–11.
42. "What Chile Thinks of Col. North," *La Libertad Electoral* (Santiago), n.d., translated and reprinted in *FN* (London), 11 February 1889; *La Libertad Electoral* (Santiago), 27/28 February 1889; *La Epoca* (Santiago), 19 March 1889; *El Ferrocarril* (Santiago), 26 May 1889.
43. Yrarrázaval, *Política económica*, pp. 22–23.
44. Brown, "Nitrate Railways," p. 468.
45. GC to AGS, Valparaíso, 10 July 1883. GMS, 11,470/7; Miller to Herbert Gibbs, Valparaíso, 9 January 1888. GMS, 11,470/10.
46. Brown, "Nitrate Railways," pp. 468, 470.
47. Blakemore, *British Nitrates*, pp. 112–13.
48. Ibid., pp. 130–34.
49. Ibid., pp. 113–14, 139–40.
50. For this argument, see Blakemore, *British Nitrates*, pp. 125–27.
51. Daubney to AGS, Valparaíso, 12 September 1890, no. 47. GMS, 11,470/12.
52. Daubney to Gibbs, Valparaíso, 16 September 1890. GMS, 11,470/12.
53. Ramírez, *Balmaceda*, pp. 77–89.
54. This practice of lobbying on behalf of foreign interests persisted after the Civil War; see Blakemore, *British Nitrates*, pp. 216–20.
55. Ibid., p. 216.
56. Kennedy to Salisbury, Viña del Mar, 25 May 1890. F.O. 16/287.
57. Billinghurst, *Capitales salitreros*, passim; AGS to GC, London, 6 September 1889. GMS, 11,471; *El Veintiuno de Mayo* (Iquique) 17 March 1887.
58. Yrarrázaval, *Política económica*, pp. 46–48; *NV*, vol. 302, 1 September 1890, fs. 225–37; Daubney to Miller, Valparaíso, 11 November 1890. GMS, 11,470/12.
59. AGS to GC, London, 5 August 1890, no. 59, 16 September 1890, no. 63. GMS, 11,471/36.
60. Blakemore, *British Nitrates*, pp. 182–83; *NV*, vol. 302, 1 September 1890, fs. 225–37. On Barros's connection to Concha y Toro and Subercaseaux, see the business transactions recorded in the notarial archives of Santiago, including: *NS*, vol. 666, 25 October 1884, fs. 428–32; *NS*, vol. 650, 29 August 1883, f. 238. On the links between Folsch, Martin, and Subercaseaux, see pp. 36 and 75, above.
61. Blakemore, *British Nitrates*, p. 162.
62. Valenzuela, *Political Brokers*, p. 202; Blakemore, *British Nitrates*, pp. 117–18. On

the composition of Balmaceda's cabinets, see Encina, *Historia de Chile*, XIX, passim.
63. Encina, *Historia de Chile*, XIX, p. 187; Valenzuela, *Political Brokers*, p. 190.
64. Encina, *Historia de Chile*, XIX, p. 203.
65. *La Patria* (Valparaíso), 3 July 1889.
66. Blakemore, *British Nitrates*, p. 124.
67. Marx, "The Eighteenth Brumaire," p. 460.
68. Karen L. Remmer, "The Timing, Pace and Sequence of Political Change in Chile, 1891–1925," *HAHR* 57 (May 1977), p. 224.
69. Valenzuela, *Political Brokers*, p. 200.
70. Blakemore, *British Nitrates*, p. 174; Valenzuela, *Political Brokers*, p. 195.
71. Valenzuela, *Political Brokers*, pp. 195–96.
72. Brown, "Nitrate Industry," p. 120.
73. In 1908 a British consul reported that Chileans controlled 38 percent of the industry. However, given the figures in Table 16 for 1906, this, no doubt, included joint Chilean-European ventures; see Brown, "Nitrate Industry," p. 120, fn. 93.
74. Verification of this hypothesis awaits a detailed study of the industry after 1891. However, my own preliminary research in this period tends to confirm it.
75. Blakemore, *British Nitrates*, pp. 216–20.
76. Marx, "The Eighteenth Brumaire," p. 460.

CHAPTER IX

1. Pinto, *Desarrollo frustrado*, p. 82.
2. Foster-Carter, "Modes of Production," passim.
3. On the European cases, see Brenner, "Capitalist Development," pp. 67–73, 75–77.
4. Fernando H. Cardoso and Enzo Faletto, *Dependencia y desarrollo en América Latina* (Mexico, 1969), pp. 42–53.
5. Göran Therborn, "The Travail of Latin American Democracy," *NLR* 113–14 (January/April 1979), pp. 99–104.
6. Wright, "Agriculture and Protectionism," p. 58.
7. Kirsch, *Industrial Development*, pp. 67–77, 157.
8. Fine, "Origins of Capitalist Development," pp. 92–95. Fine postulates a major role for labor in the formation of the social consensus for reform. The failure of Chilean labor in this regard would be a most fruitful area for investigation.
9. On the fragmentation of the labor force, see Johnson, "Internal Migration," pp. 442–63.
10. On these two perspectives, see Brenner, "Capitalist Development," passim.
11. On these developments after World War I, see Mamalakis, *Chilean Economy*, pp. 40–45, 66–67; Cardoso and Faletto, *Dependencia*, p. 134; Alan Angell, *Politics and the Labour Movement in Chile* (London, 1972), pp. 11–81; Brian Loveman, *Struggle in the Countryside: Politics and Rural Labor in Chile, 1919–1973*

(Bloomington, Ind., 1976), pp. 200–203; Constantine C. Menges, "Public Policy and Organized Business in Chile: A Preliminary Analysis," *Journal of International Affairs* 20 (1966), pp. 343–65. On the functions and problems of the state under peripheral capitalism, see W. Ziemann and M. Lanzendörfer, "The State in Peripheral Societies," *Socialist Register* (1977), pp. 160–67.

APPENDIX II

1. Brown, "Nitrate Industry," p. 196; Bauer, *Rural Society*, p. 156. The assumed 100 percent markup would clearly be an exaggeration, since company store profits averaged only 30 percent before 1910. Furthermore, store prices were often reduced in order to attract workers in the nitrate zone's highly competitive labor market; see Stickell, "Migration and Mining," pp. 296–97.
2. Bauer, *Rural Society*, pp. 155–59. On the low level of wages in Chile compared with those in the developed countries, see Chapter IV, note 39, above.
3. Compañía de Salitres y Ferrocarril de Antofagasta, *Memorias* 1–3 (Valparaíso, 1872–73); Bohl to Hayne, Valparaíso, 16 May 1873. GMS, 11,121; Smail to Read, Oficina Limeña, 3 February 1879. GMS, 11,472/11. On the higher real wages earned by nitrate workers compared with laborers in other sectors of the economy after 1880, see Stickell, "Migration and Mining," pp. 259–71.
4. *La Industria* (Iquique), 31 January 1888; *PP*, 1889, 78, "Report on the Trade and Commerce of Coquimbo for the Years 1887–1888," p.1. Similar reports throughout the decade indicate this was not an isolated instance; see, for example, *PP*, 1886, 66, "Report by Consul Grierson on the Trade and Commercce of Coquimbo for the Year 1885," pp. 628–29. For a discussion of the problem in subsequent decades, see Alfred A. Winslow, "Scarcity of Laborers in Chile, 20 February 1913," and J. Perkins, "Labour Situation in Northern Chile, 8 September 1921." General Records of the Department of State, Record Group 59, National Archives, Washington, D.C. For a detailed discussion of labor scarcity after 1880, see Stickell, "Migration and Mining," pp. 47–81.
5. Ann Louise Hagerman Johnson in her study of Chilean migration patterns concludes that migration to the nitrate region after 1880 did not disrupt the countryside because labor demands of agriculture in the Central Valley increased only slightly under conditions of an acute labor surplus. Yet she also notes the limits of the wage market, pointing out that the migrations did not uproot settled peasants, affecting primarily those rural elements which had already been exposed to wage labor in public works projects and other such endeavors; see Ann Louise Hagerman Johnson, "Internal Migration," pp. 355–67.

Sources

I. ARCHIVAL SOURCES

1. Archivo Nacional, Santiago de Chile

Archivo del Ministerio de Hacienda. vols. 1089, 1091, 1043, 1138, 1240, 1241, 1244, 1338–1340, 1363, 1364, 1390–1392, 1493, 1494, 1551, 1585, 1669–1671, 1787–1794, 1913, 1939–1942, 2015, 2060–2065, 2113, 2114, 2141, 2245, 2321–2324.
Archivo del Ministerio de Relaciones Esteriores. Vol. 176.
Archivos Judiciales de Iquique. 1870–1890.
Archivos Notariales de Iquique. Vols. 1–128. 1864–1891.
Archivos Notariales de Santiago. Vols. 620–797. 1881–1890.
Archivos Notariales de Tarapacá. Vols. 8–14. 1867–1894.
Archivos Notariales de Valparaíso. Vols. 158–314. 1870–1891.
Fondo Nuevo "Varios." Vols. 413–414.

2. National Archives, Washington, D.C.

General Records of the Department of State, Record Group 59.

3. Public Record Office, London, England

Foreign Office Archives. Embassy and Consular Archives, Correspondence—Peru (F.O. 177). Vols. 121–185. 1869–1886.
Foreign Office Archives. General Correspondence—Chile (F.O. 16). Vols. 163–288. 1870–1893.
Foreign Office Archives. General Correspondence—Peru (F.O. 61). Vols. 258–387. 1870–1890.

4. Guildhall Library, London, England

Archives of Antony Gibbs & Sons. MS nos. 11,033; 11,033A/1–8; 11,040; 11,049A; 11,067; 11,120; 11,121; 11,122; 11,123; 11,128; 11,129; 11,130; 11,131; 11,132; 11,135; 11,138/3; 11,470/1–12; 11,471/1–39; 11,472/1–4.

II. PRINTED SOURCES

1. Books and Articles

Aldunate Solar, Carlos. *Leyes, decretos i documentos relativos a salitreras.* Santiago, 1907.
Allen, Arthur W. *The Recovery of Nitrate from Chilean Caliche.* Philadelphia, 1921.
Amin, Samir. *Unequal Development.* Translated by Brian Pearce. New York, 1976.
Angell, Alan. *Politics and the Labour Movement in Chile.* London, 1972.
Arce R. Issac. *Narraciones históricas de Antofagasta.* Antofagasta, 1930.
Astorquiza, Octavio, comp. and ed. *Lota: antecedentes históricos, con una monografía de la Compañía Minera e Industrial de Chile.* Concepción, 1929.
Balmori, Diana, and Oppenheimer, Robert. "Family Clusters: Generational Nucleation in Nineteenth Century Argentina and Chile." *Comparative Studies in Society and History* 21 (April 1979), pp. 231–61.
Bañados Espinosa, Julio. *Balmaceda: su gobierno y la revolución de 1891.* 2 vols. Paris, 1894.
Banco Mobilario. *Memoria del Banco Mobilario.* Santiago, 1877.
Basarde, Jorge. *Historia de la república del Perú.* 6 vols. 5th ed., rev. and enl. Lima, 1961–1962.
Bauer, Arnold J. *Chilean Rural Society from the Spanish Conquest to 1930.* Cambridge, England, 1975.
———. Review of *The Heroic Image in Chile: Arturo Prat, Secular Saint,* in *Journal of Latin American Studies* 7 (May 1975), pp. 159–61.
Bermúdez Miral, Oscar. "El salitre de Tarapacá y Antofagasta durante la ocupación militar chilena." *Anales de la Universidad del Norte* (Antofagasta), 5 (1966), pp. 131–82.
———. *Historia del salitre desde sus orígenes hasta la Guerra del Pacífico.* Santiago, 1963.
Billinghurst, Guillermo E. *Los capitales salitreros de Tarapacá.* Santiago, 1889.
Blakemore, Harold. *British Nitrates and Chilean Politics, 1886–1896: Balmaceda & North.* London, 1974.
———. "Limitations of Dependency: an Historian's View and Case Study." *Boletín de Estudios Latinoamericanos y del Caribe* 18 (June 1975), pp. 74–87.
———. "The Chilean Revolution of 1891 and Its Historiography." *Hispanic*

American Historical Review 45 (August 1965), pp. 393–421.
Blest Gana, Alberto. "La situación financiera de Chile en 1878." *Revista Chilena* 13 (April 1922), pp. 485–94.
Bonilla, Heraclio. *Guano y burguesía en el Perú*. Lima, 1974.
Bowman, Isaiah. *Desert Trails of Atacama*. New York, 1924.
Brenner, Robert. "The Origins of Capitalist Development: A Critique of Neo-Smithian Marxism." *New Left Review* 104 (July–August 1977), pp. 25–92.
Brown, J. R. "Nitrate Crises, Combinations, and the Chilean Government in the Nitrate Age." *Hispanic American Historical Review* 43 (May 1963), pp. 230–46.
———. "The Chilean Nitrate Railways Controversy." *Hispanic American Historical Review* 38 (November 1958), pp. 465–81.
———. "The Frustration of Chile's Nitrate Imperialism." *Pacific Historical Review* 32 (November 1963), pp. 383–96.
Bulnes, Gonzalo. *Guerra del Pacífico*. 3 vols. Valparaíso, 1911–19.
Burr, Robert N. *By Reason or Force: Chile and the Balancing of Power in South America, 1830–1905*. Berkeley, 1965.
Cairncross, A. K. *Home and Foreign Investment, 1870–1913*. Cambridge, England, 1953.
Cardoso, Fernando H., and Faletto, Enzo. *Dependencia y desarrollo en América Latina*. Mexico, 1969.
Castle, W. M. F. *Sketch of the City of Iquique, Chili, South America during Fifty Years*. Plymouth, England, 1887.
Centner, C. W. "Great Britain and Chilean Mining, 1830–1914." *Economic History Review* 12 (1942), pp. 76–82.
Chile. Comisión Consultiva de Salitres. *Informe que la comisión consultiva de salitres presenta al Señor Ministro de Hacienda*. Santiago, 1880.
Chile. Congreso. *Boletín de las sesiones estraordinarias de la cámara de senadores*. Santiago, 1880–90.
Chile. Congreso. *Boletín de las sesiones ordinarias de la cámara de diputados*. 1879–90.
Chile. Congreso. *Boletín de las sesiones ordinarias de la cámara de senadores*. Santiago, 1879–90.
Chile. Ministerio de Hacienda. *Memorias del Ministro de Hacienda presentada al congreso nacional*. Santiago, 1880–92.
Chile. Ministerio de Hacienda. Sección Salitre. *Antecedentes sobre la industria salitrera*. Santiago, 1925.
Chile. Ministerio del Interior. *Memoria del Intendente de Tarapacá correspondiente a 1886*. Santiago, 1887.

———. *Memoria del Intendente de Tarapacá correspondiente a 1887.* Santiago, 1888.

———. *Memoria del Intendente de Tarapacá presentada al Señor Ministro del Interior en 1889.* Santiago, 1889.

Compañía Beneficiadora de Oro de Quillota. *Estatutos de la Compañía Beneficiadora de Oro de Quillota.* Valparaíso, 1881.

Compañía Comercial de Remolcadores. *Estatutos de la Compañía Comercial de Remolcadores.* Valparaíso, 1882.

Compañía de Gas de Concepción. *Estatutos de la Compañía de Gas de Concepción.* Valparaíso, 1887.

Compañía de Maderas y Carbón. Estatutos de la Compañía de Maderas y Carbón. Valparaíso, 1883.

Compañía de Salitres y Ferrocarril de Antofagasta. *Memorias de la Compañía de Salitres y Ferrocarril de Antofagasta.* Valparaíso, 1873–91.

Compañía Huanchaca de Bolivia. *7a Memoria de la Compañía Huanchaca de Bolivia.* Valparaíso, 1877.

———. *18a Memoria de la Compañía Huanchaca de Bolivia.* Valparaíso, 1890.

———. *20a Memoria de la Compañía Huanchaca de Bolivia.* Valparaíso, 1892.

Compañía Refinería de Azucar de Viña del Mar. *Estatutos de la Compañía Refinería de Azucar de Viña del Mar.* Valparaíso, 1887.

Compañía Salitrera America. *Estatutos de la Compañía Salitrera America.* Valparaíso, 1873.

Compañía Salitrera Negreiros. *Estatutos de la Compañía Salitrera Negreiros.* Valparaíso, 1872.

Compañía Salitrera Pisagua. *Memoria de la Compañía Salitrera Pisagua.* Valparaíso, 1872.

Compañía Salitrera Sacramento. *Estatutos de la Compañía Salitrera Sacramento.* Valparaíso, 1872.

Compañía Salitrera San Carlos. *Estatutos de la Compañía Salitrera San Carlos.* Valparaíso, 1873.

Compañía Salitrera Valparaíso. *Estatutos de la Compañía Salitrera Valparaíso.* Valparaíso, 1873.

———. *Memoria de la Compañía Salitrera Valparaíso.* Valparaíso, 1877.

Compañía Telegrafo Transandino. *Estatutos de la Compañía Telegrafo Transandino.* Valparaíso, 1887.

Cortés Lía, and Fuentes, Jordi. *Diccionario político de Chile, 1810–1966.* Santiago, 1967.

Cruchaga, Miguel. *Salitre y guano.* Madrid, 1929.

Dancuart, P. Emilio, ed. *Anales de la hacienda pública del Perú. Historia y legislación fiscal de la república.* 22 vols. Lima, 1902–8.

Dennis, William Jefferson. *Tacna and Arica: an Account of the Chile–Peru Boundary Dispute and of the Arbitrations of the United States.* New Haven, Conn., 1931; reprint edition, Hamden, Conn., 1967.

Donald, M. B. "History of the Chile Nitrate Industry." *Annals of Science* 1 (January 1936), pp. 29–47 (April 1936), pp. 193–216.

Duffield, A. J. *Peru in the Guano Age.* London, 1877.

Encina, Francisco A. *Historia de Chile desde la prehistoria hasta 1891.* 20 vols. 2d ed. Santiago, 1955.

———. *Nuestra inferioridad económica: sus causas, sus consecuencias.* Santiago, 1912.

Escobar Cerda, Luis. *El mercado de valores.* Santiago, 1959.

Espinoza, Enrique. *Jeografía descriptiva de la república de Chile.* 4th ed., enl. Santiago, 1897.

Fábrica Nacional de Cerveza. *la Memoria de la Fábrica Nacional de Cerveza.* Valparaíso, 1890.

Ferrocarril Urbano de Santiago. *16a Memoria del Ferrocarril Urbano de Santiago.* Santiago, 1882.

Fetter, Frank Whitson. *Monetary Inflation in Chile.* Princeton, 1931.

Fifer, J. Valerie. *Bolivia: Land, Location and Politics since 1825.* Cambridge, England, 1972.

Figueroa, Virgilio. *Diccionario histórico y biográfico de Chile.* 5 vols. Santiago, 1926–35.

Fine, Ben. "On the Origins of Capitalist Development," *New Left Review* 109 (May–June 1978), pp. 88–95.

Foster-Carter, Aidan. "The Modes of Production Controversy." *New Left Review* 107 (January–February 1978), pp. 47–77.

Frank, Andre Gunder. *Capitalism and Underdevelopment in Latin America: Historical Studies of Chile and Brazil.* New York, 1967.

Furtado, Celso. *Economic Development of Latin America: A Survey from Colonial Times to the Cuban Revolution.* Translated by Suzette Macedo. Cambridge, England, 1970.

Gana, Henry Sewell. *British Capital and Chilian Industry, Nitrates, Gold Mines and Coal Mines.* London, 1889.

Great Britain. Parliament. *Parliamentary Papers.* 1874, vol. 66; 1876, vols. 73, 75; 1877, vol. 81; 1878, vols. 72, 75; 1880, vol. 74; 1881, vol. 91; 1882, vols. 70, 81; 1883, vols. 72, 73; 1885, vols. 77, 79; 1886, vol. 66; 1887, vol. 83; 1888, vol. 100; 1889, vols. 77, 78; 1890, vols. 73, 74.

Greenhill, Robert G., and Miller, Rory M. "The Peruvian Government and the Nitrate Trade, 1873–1879." *Journal of Latin American Studies* 5 (May 1973), pp. 107–31.

Heise González, Julio. *Historia de Chile, el periodo parlamentario, 1861–1925.* Santiago, 1974.

Hernández Cornejo, Roberto. *El salitre: resumen histórico desde su descubrimiento y explotación.* Valparaíso, 1930.

Herrmann, Alberto. *La producción de oro, plata i cobre en Chile desde los primeros dias de la conquista hasta fines de agosto de 1894.* Santiago, 1894.

Hobsbawm, E. J. *The Age of Capital, 1848–1875.* New York, 1975.

Jobet Búrquez, Julio César. *Ensayo crítico del desarrollo económico-social de Chile.* Santiago, 1955.

Jones, Charles. "Insurance Companies." In *Business Imperialism, 1840–1930: An Inquiry Based on British Experience,* pp. 53–74. Edited by D. C. M. Platt. Oxford, 1977.

Joslin, David. *A Century of Banking in Latin America.* London, 1963.

Kaempffer, Enrique. *La industria del salitre y del yodo con ilustraciones y un glosario de voces técnicas.* Santiago, 1914.

Kaplan, Marcos, *Formación del estado nacional en América Latina.* Santiago, 1969.

Kay Geoffrey. *Development and Underdevelopment: A Marxist Analysis.* New York, 1975.

Kirsh, Henry W. *Industrial Development in a Traditional Society.* Gainesville, Florida, 1977.

Kling, Merle. "Toward a Theory of Power and Political Instability in Latin America." Reprinted from *Western Political Quarterly,* 9 (March 1956), pp. 21-35, In *Latin America: Reform or Revolution,* pp. 76-93. Edited by James Petras and Maurice Zeitlin. New York, 1968.

La Feber, Walter, *The New Empire: An Interpretation of American Expansion, 1860–1898.* Ithaca, N.Y., 1971.

Landes, David S. *The Unbound Prometheus.* Cambridge, England, 1969.

La Protectora Compañía Chilena de Seguros. *Estatutos de la Protectora Compañía Chilena de Seguros.* Valparaíso, 1886.

Le Feuvre, René F., and Dagnino, Arturo. *El salitre de Chile ó nitrato de soda.* Santiago, 1893.

Levin, Jonathan V. *The Export Economies: Their Pattern of Development in Historical Perspective.* Cambridge, Mass., 1960.

Loveman, Brian. *Chile: The Legacy of Hispanic Capitalism.* New York, 1979.

———. *Struggle in the Countryside: Politics and Rural Labor in Chile, 1919–1973.* Bloomington, Ind., 1976.

Lubbock, Basil. *The Nitrate Clippers.* Glasgow, 1932.

McQueen, Charles A. *Peruvian Public Finance.* Department of Commerce, Bureau of Foreign and Domestic Commerce, Trade and Promotion Series No. 30. Washington, D.C., 1926.

Mamalakis, Markos J. *The Growth and Structure of the Chilean Economy: From Independence to Allende.* New Haven, Conn., 1976.

———. "The Role of Government in the Resource Transfer and Resource Allocation Processes: the Chilean Nitrate Sector, 1880–1930." In *Government and Economic Development,* pp. 181–210. Edited by Gustav Ranis. New Haven, Conn., 1971.

Mandel, Ernest. *Late Capitalism.* Translated by Joris de Bres. London, 1979.

Marín Vicuña, Santiago. *Los ferrocarriles de Chile.* 3d ed. Santiago, 1912.

Martner, Daniel. *Estudio de política comercial chilena e historia económica nacional.* 2 vols. Santiago, 1923.

Marx, Karl. *Capital: A Critique of Political Economy.* 3 vols. Translated by Ernest Untermann, edited by Frederick Engels. Chicago, 1909.

———. "The Eighteenth Brumaire of Louis Bonaparte." In *The Marx-Engels Reader,* pp. 436–525. Edited by Robert C. Tucker. New York, 1972.

Matthew, W. M. "Foreign Contractors and the Peruvian Government at the Outset of the Guano Trade." *Hispanic American Historical Review* 52 (November 1972), pp. 598–620.

———. "The Imperialism of Free Trade: Peru, 1820–70." *Economic History Review* 2d ser., 21 (December 1968), pp. 562–77.

———. "Peru and the British Guano Market, 1840–1870." *Economic History Review* 2d ser., 23 (April 1970), pp. 112–28.

Maude, Wilfred. *Antony Gibbs & Sons Ltd. Merchants and Bankers, 1808–1958.* London,, 1958.

Menges, Constantine C. "Public Policy and Organized Business in Chile: A Preliminary Analysis." *Journal of International Affairs* 20 (1966), pp. 343–65.

Millington, Herbert. *American Diplomacy and the War of the Pacific.* New York, 1948.

Mingay, G. E. "The Transformation of Agriculture." In *The Long Debate on Poverty,* pp. 25–60. R. M. Hartwell et.al. London, 1972.

Monteón, Michael. "The British in the Atacama Desert: The Cultural Bases of Economic Imperialism." *Journal of Economic History* 35 (March 1975), pp. 117–33.

Moore, Barrington, Jr. *Social Origins of Dictatorship and Democracy: Lord and Peasant in the Making of the Modern World.* Boston, 1967.

O'Brien, Philip J. "A Critique of Latin American Theories of Dependency." In *Beyond the Sociology of Development: Economy and Society in Latin America and Africa*, pp. 7–27. Edited by Ivar Oxaal, Tony Barnett, and David Booth. London, 1975.

O'Brien, Thomas F. "The Antofagasta Company: A Case Study of Peripheral Capitalism." *Hispanic American Historical Review* 60 (February, 1980), pp. 1–31.

Ossa Bourne, Samuel. "Don José Santos Ossa." *Revista Chilena de Historia y Geografía* 67 (1931), pp. 43–90, 68 (1931), pp. 112–41, 69 (1931), pp. 186–215, 72 (1932), pp. 176–228.

Pederson, Leland R. *The Mining Industry of the Norte Chico, Chile*. Evanston, Ill., 1966.

Pike, Frederick B. "Aspects of Class Relations in Chile, 1850–1960." *Hispanic American Historical Review* 43 (February 1963), pp. 14–33.

———. *Chile and the United States, 1880–1962: The Emergence of Chile's Social Crisis and the Challenge to United States Diplomacy*. Notre Dame, Ind., 1963.

Pinto Santa Cruz, Aníbal. *Chile, un caso de desarrollo frustrado*. 3d ed. Santiago, 1973.

Platt, D. C. M. *Latin America and British Trade, 1806–1914*. London, 1972.

Ramírez Necochea, Hernán. *Balmaceda y la contrarrevolución de 1891*. 3d ed., rev. and enl. Santiago, 1972.

———. *La guerra civil de 1891: antecedentes económicos*. Santiago, 1951.

Remmer, Karen L. "The Timing, Pace and Sequence of Political Change in Chile, 1891–1925." *Hispanic American Historical Review* 57 (May 1977), pp. 205–30.

Resúmen de la hacienda pública de Chile desde 1833 hasta 1914. London, 1914.

Reyes N., Enrique. *El desarrollo de la conciencia proletaria en Chile (el ciclo salitrero)*. Santiago, 1973.

Rippy, J. Fred. "The British Investment 'Boom' of the 1880's in Latin America." *Hispanic American Historical Review* 29 (May 1949), pp. 281–86.

———. "British Investment in the Chilean Nitrate Industry." *Inter-American Economic Affairs* 8 (Autumn 1954), pp. 3–10.

———. "Economic Enterprises of the 'Nitrate King' and His Associates in Chile." *Pacific Historical Review* 17 (November 1948), pp. 457–65.

Romero, Emilio. *Historia económica del Perú*. 2 vols. 2d ed. Lima, 1968.

Rondizzoni, F. *Minerales, guano i salitre de Atacama. Medidas oficiales para el fomento de la industria*. Santiago, 1877.

Ross Agustín. *Chile, 1851–1910. Sesenta años de cuestiones monetarias y fin-*

ancieras y de problemas bancarias. Santiago, 1911.
Russell, William Howard. *A Visit to Chile and the Nitrate Fields of Tarapacá.* London, 1890.
Salas Edwards, Ricardo. *Balmaceda y el parlamentarismo en Chile: un estudio de psicología política chilena.* 2 vols. Santiago, 1914–1925.
San Cristóval, Evaristo. *Manuel Pardo y Lavalle, su vida y su obra.* Lima, 1945.
Santelices, Ramón E. *Los bancos chilenos.* Santiago, 1893.
Sater, William F. "Chile and the World Depression of the 1870's." *Journal of Latin American Studies* 11 (May 1979), pp. 67–99.
———. "Economic Nationalism and Tax Reform in Late Nineteenth Century Chile." *The Americas* 33 (October 1976), pp. 311–35.
Saul, S. B. *Studies in British Overseas Trade, 1870–1914.* Liverpool, 1960.
Segall, Marcelo. *Desarrollo del capitalismo en Chile: cinco ensayos dialécticos.* Santiago, 1953.
Semper, E., and Michels, E. *La industria del salitre en Chile.* Translated and augmented by Javier Gandarillas and Orlando Ghigliotto Salas. Santiago, 1908.
Soto Rojas, Salvador. *Las riquezas de Chile en sus industrias y comercio.* Santiago, 1906.
Stein, Stanley J., and Stein, Barbara H. *The Colonial Heritage of Latin America. Essays on Economic Dependence in Perspective.* New York, 1970.
Stewart, Watt. *Henry Meiggs, Yankee Pizarro.* Durham, N.C., 1946.
Stone, Irving. "British Long-Term Investment in Latin America, 1865–1913." *Business History Review* 42 (Autumn 1968), pp. 311–39.
Subercaseaux, Guillermo. *Monetary and Banking Policy of Chile.* Oxford, 1922.
Sweezy, Paul M. *The Theory of Capitalist Development.* New York, 1968.
Therborn, Göran, "The Travail of Latin American Democracy." *New Left Review* 113–114 (January–April 1979), pp. 71–109.
U.S. Department of State. *Papers Relating to the Foreign Relations of the United States Transmitted with the Annual Message of the President to Congress.* Washington, D. C., 1876.
Urzúa Valenzuela, Germán. *Evolución de la administración pública chilena (1818–1968).* Santiago, 1970.
Valdés Vergara, Francisco. *La crisis salitrera i las medidas que se proponen para remediarla.* Santiago, 1884.
———. *Memoria sobre la administración de Tarapacá presentada al supremo gobierno.* Santiago, 1884.
———. *Problemas económicos de Chile.* Valparaíso, 1913.

Valenzuela, Arturo. *Political Brokers in Chile: Local Government in a Centralized Polity.* Durham N.C., 1977.
Véliz, Claudio. *Historia de la marina mercante de Chile.* Santiago, 1961.
Vicuña Mackenna, Benjamín. *El libro del cobre i del carbón de piedra en Chile.* Santiago, 1883.
Vidal Gormaz, Francisco. *Estudio sobre el puerto de Iquique.* Santiago, 1880.
Walker Martínez, Carlos. *Historia de la administración Santa María.* 2 vols. Santiago, 1888–89.
Weaver, Striton F. "Growth Theory and Chile—The Problem of Generalizing from Historical Example." *Journal of Inter-American Studies and World Affairs* 12 (January 1970), pp. 55–61.
Woll, Allen L. "For God or Country: History Textbooks and the Secularization of Chilean Society, 1840–1870." *Journal of Latin American Studies* 7 (May 1975), pp. 23–43.
Wright, Thomas C. "Agriculture and Protectionism in Chile, 1880–1930." *Journal of Latin American Studies* 7 (May 1975), pp. 45–58.
———. "The Politics of Inflation in Chile, 1888–1918." *Hispanic American Historical Review* 53 (May 1973), pp. 239–59.
Yrarrázaval Larraín, José Miguel. *El Presidente Balmaceda.* 2 vols. Santiago, 1940.
———. *La política económica del presidente Balmaceda.* Santiago, 1963.
Zegers A., Cristián. *Aníbal Pinto: historia política de su gobierno.* Santiago, 1969.
Ziemann, W., and Lanzendörfer, M. "The State in Peripheral Societies." *Socialist Register* (1977), pp. 143–77.

2. Unpublished Works

Bader, Thomas, McLeod. "A Willingness to War: A Portrait of the Republic of Chile During the Years Preceding the War of the Pacific." Ph.D. diss., University of California, Los Angeles, 1967.
Brown, Joseph Robert. "The Chilean Nitrate Industry in the Nineteenth Century." Ph.D. diss., Louisiana State University, 1954.
Johnson, Ann Louise Hagerman. "Internal Migration in Chile to 1920: Its Relationship to the Labor Market, Agricultural Growth and Urbanization." Ph.D. diss., University of California, Davis, 1978.
Marcella, Gabriel. "The Structure of Politics in Nineteenth Century Spanish America: the Chilean Oligarchy, 1833–1891." Ph.D. diss., University of Notre Dame, 1973.

Stickell, Arthur Lawrence. "Migration and Mining: Labor in Northern Chile in the Nitrate Era, 1880–1930." Ph.D. diss., Indiana University, 1979.

3. Periodicals

Brazil and River Plate Mail (London), 1873.
Burdetts Official Intelligence (London), vols. 4–11, 1886–1893.
Economist (London), 1886–1891.
El Ferrocarril (Santiago), 1880–1881, 1889–1890.
El Heraldo (Valparaíso), 1889.
El Mercurio (Valparaíso), 1870–1890.
El Tarapacá (Iquique), 1884–1887.
El Veintiuno de Mayo (Iquique), 1880–1888.
Financial News (London), 1886–1891.
La Epoca (Santiago), 1884, 1889–1890.
La Industria (Iquique), 1882–1890.
La Libertad Electoral (Santiago), 1889–1890.
La Patria (Valparaíso), 1873–1875, 1879–1890.
Scientific American (New York), 1880–1890.
South American Journal (London), 1884–1891.
South Pacific Times (Callao), 1872.
Statist (London), 1886–1891.
Valparaíso and West Coast Mail, 1873–1875.

Index

Agriculture: exports, 2-3, 8; and world trade, 2; wages in, 11, 158-59; class relations in, 11; labor intensive, 24-25, 80; links to nitrate industry, 78-80

Aguas Blancas: Chilean nitrate enterprises in, 34; encouragement of nitrate industry in, 46-47; congressional debate on nitrate duty, 58-60

Aldunate, Luis, 55

Altamirano, Eulogio, Senator, 138

Amunategui, Miguel, 51, 53

Ancón, Treaty of, 64

Anglo-Peruvian Bank, 34-36

Antofagasta (city), 49

Antofagasta (province), 56

Antofagasta and Bolivia Railway Company, 107

Antofagasta Nitrate and Railway Company: and Gibbs, 18; early development, 22-25; and Peruvian expropriation, 32; taxes on, 49, 56-57, 60

Araucanian frontier, 102, 106

Arteaga Alemparte, Justo, 58-59

Associated Banks, The, 29-31

Atacama Desert: location, 5; opportunities for nitrate enterprises, 36; encouragement of nitrate industry in, 46-47; Chilean penetration of, 48-49

Avery Hill, 114

Balmaceda, José Manuel, President: government's expenditures, 81-82; warning on nitrate monopoly, 111; career of, 126-28; nitrate policy of, 133-37

Balta, José, President, 20

Banco de Valparaíso: branch in Iquique, 57; and John Thomas North, 67; loan to Clark, Eck and Company, 71; financing of nitrates and profits, 83-84; and First Combination, 90

Banco Garantizador de Valores, 106-07

Banco Mobilario, 83-84, 140

Banco Nacional de Chile: branch in Iquique, 57; credit to Jenaro Canelo, 75; financing nitrates and profits, 83-84; and nitrate industry, 90

Banco Nacional del Perú, 14

Bancos Asociados, Los See Associated Banks, The

Bankers, Chilean, 3

Bank of London, Mexico and South America, 34, 39, 54

Bank of Tarapacá and London, 117, 135, 145

Banks, Chilean: first organized, 2; and the crisis of 1878, 47; share in nitrate revenues, 129, 132-33

Barros, Lauro, 139, 141

Barros Luco, Ramón, 56-57

Bauer, Arnold, 3, 11, 44

205

INDEX

Bernstein, Julio, 104
Billinghurst, Guillermo, 139-40
Blaine, James, 52
Blest Gana, Alberto: views on crisis of 1878, 47-48, and Peruvian bondholders, 50; and Gibbs's nitrate claim, 53-54; views on public employment, 130
Boletín do la Sociedad Nacional de Agricultura, 56, 90-91
Bolivia: policy toward Antofagasta Nitrate and Railway Company, 32, 49; and War of the Pacific, 42-49
Bondholders, Peruvian, 50-51, 61
Bordes, Antoine D., 18, 104
Bulnes, Gonzalo: founder of Ceballos, Sanz y Compañía, 75; report on J. D. Campbell and Company, 91; his critique of Balmaceda's nitrate policy, 134
Bureaucracy, Chilean: expansion of, 4, 82, 125, 129-30, 133, 144; and Chamber of Deputies, 128

Caja de Credíto Hipotecario, 44, 71, 131
Caleta Buena, 71, 91, 101
Caliche: description of, 5; declining grades of, 12, 68, 121
Campbell, J. D. and Company: founding of, 16; purchase of *oficina* Agua Santa, 23; joins Peruvian expropriation, 35; investment in *oficina* Agua Santa, 38; refusal to pay nitrate duty, 52; position in industry, 65; development of Shanks system, 69-70; capital, 71; interest in Herrera monorail, 101
Canelo, Jenaro, 14, 34, 74-75
Capellanías, 44
Capital centralization, 112, 119, 122-23
Capital concentration, 70-73, 86
Capital, industrial, 25
Capital market, British, 72, 112-13
Capital market, Chilean, 3, 10, 12, 79, 81, 83-84, 94
Capital market, European, 35
Capital, merchant, 25
Capitalism, European: penetration of Peruvian economy, 27; commercial interchange with periphery, 43; relationship to Chilean economy, 62, 94, 109, 151-52

Capitalism, monopoly, 112-13, 123, 145, 151-52
Caracoles, 10
Cardoso, Fernando H., 151
Carrizal railway, 66
Catholic church, 125-26
Censos, 44
Central Valley, 11
Chamber of Deputies, 58-59, 128
Chambre La, Gautreau and Company, 13, 22, 34
Chañaral railroad, 131-32
Chañarcillo, 2
Churchill, Randolph, 114
Civil War of 1891, 124, 143-46, 152
Clark, Eck and Company: founding of, 16-17; loan repayments, 23; obtains production contract, 36; position in nitrate industry, 65; bank loans, 71
Clark, Melbourne, 15
Class relations, precapitalist, 11, 81, 147, 150, 155. *See also* Social productive relations
Club de la Unión, 98
Cobija, 10
Combination, First, 73, 90, 92, 111-12
Combination, Second, 145
Compañía Chileana de Consignaciones y Depósitos, 10, 13, 22
Compañía Comercial de Remolcadores, 103
Compañía de Gas de Concepción, 102-03
Compañía de Lota, 2
Compañía de Refineria de Azucar de Viña del Mar, 104
Compañía de Salitres de Tarapacá *See* T.N.C.
Compañía de Salitres y Ferrocarril de Agua Santa, 139-40
Compañía Huanchaca de Bolivia, 107
Concha y Toro family, 57
Concha y Toro, Melchor: director of Compañía Salitrera San Carlos, 11; views on War of Pacific, 49; views on nitrate duty, 56, 59-61; interview with Gibbs, 90; biography, 98-99, 107-08; political affiliation, 141
Conservative party, 126
Constitution of 1833, 124-25
Control mechanisms, non-wage, 96
Copper prices, 37, 45-46, 86

INDEX 207

Copper production, 2, 43, 86
Cornejo y Compañía, 10
Corredor, 83
Costa, Andrés, 183n3
Costa, Antonio, 97-100
Costa, Emeterio, 98
Council of State, 136
Courcelle-Seneuil, 47
Crisis of 1878, 93, 145

Daubney, Henry Giles, 138
Dawson, John, 67, 140
Délano, Eduardo, 11, 74-75, 98-99, 103-04
Delegación Fiscal de Salitreras y Guaneras, 133
Dependency relationships, 21, 41, 149, 151-55
Dependent development, 147-49
Dissidents *See* Independent Liberals

Earthquake of 1868, 9, 16
Echeverria family, 87
Economy, Chilean: colonial structures, 2; exports, 3, 5, 46, 153-54; compared to Peruvian economy, 7; growth of, 24, 37; limits imposed by, 41; oscillations of, 45; crisis of 1878 in, 46-47; foreign debt, 46; decline of, 57
Economy, Peruvian: external debt, 20; development of, to 1860s, 26-27; collapse, 35; structural problems of, 40-41
Education, 81-82
Edwards, A. y Compañía, 13, 22
Edwards, clan, 3
Edwards Ossandón, Agustín: partnership in Antofagasta Company, 18; response to Peruvian expropriation, 32; death of, 32; control of copper trade, 43
Edwards Ross, Agustín: ownership of nitrate certificates, 36; president of Antofagasta Company, 49; biography, 97-99, 104-05; ownership of Chañaral railroad, 132, political affiliation, 134
Elite, Chilean: description of, 3; attitude toward, nitrate duty, 60-61; relationship to landed estates, 80; need to preserve premodern class relations, 87; factors determining status, 98; economic ties to nitrate industry, 150. *See also* Oligarchy

Encina, Francisco, 128
Epoca, La, 84, 135
Errázuriz, Federico, President, 45
Errázuriz, Maximiano, 56
Estaca, 7
Estanco, 21, 24
Evarts, Charles, 52
Exchange rates: pesos, 86, 160; soles, 160
Expropriation, Peruvian: development of, 26-33; promulgation of, 28; halted by President Pardo, 35; renewed by President Prado, 35; dismantled, 63; effects, 149

Ferrocarril, El, 135-36
Ficha system, 12, 92-93, 121
Financial News, 116, 118
Folsch and Martin: founding of, 16-17; halt nitrate production, 23; agreements with Francisco Subercaseaux, 36, 56, 75; position in nitrate industry, 65; links to Banco Mobilario, 90; support of Balmaceda, 140
Foreign Office: support of claims by Peruvian bondholders, 50-51; protest of Chilean seizure of Gobb's nitrate, 53; abandons bondholders' claim, 55; protests on behalf of Nitrate Railways Company, 136-37
Fundo, 11

Gallo, Angel Custodio, 8
Gamboni Vera, Pedro, 8-9
Garcia-Huidobro family, 87
Gibbs, Antony and Sons Ltd.: merchants in Chile, 3; William Gibbs and Company as a branch of, 15; involvement in guano trade, 15; financing, 16; efficiency of *oficinas*, 17-18; partnership in Antofagasta Company, 18, 23, 25, 31-32; considers closing T.N.C., 28; involvement in nitrate consignment, 30-32; nitrate production contract, 38; and nitrate certificates, 39; refusal to pay nitrate duty, 52; seizure of nitrate and protest, 53-54; position in industry, 65; rents *oficina* La Palma, buys nitrate grounds Alianza, 65; loan to John Thomas North, 67; capital, 70; bank accounts in Chile, 83; interview with Melchor Concha y Toro, 90; sale of *oficinas*, 116; stock manipulation, 118

Gibbs, Herbert, 121
Gildemeister, Henry, 31-33
Gildemeister, J. D. and Company: founding of, 16; nitrate production contract, 38; refusal to pay nitrate duty, 52; position in nitrate industry, 65; capital of, 71; sale of *oficinas*, 116
González, Marcial, 36, 55
Government revenues: increase in, 4; and customs receipts, 44; strain on, 46-47; and nitrate duty, 81, increase in, and effect on economy, 128-31
Granadino, Marcos y Hermanos, 14
Gremio de jornaleros, 91, 93, 121
Guano: on Chincha Islands, 7; competition with nitrate, 7; decline of deposits, 8-9; decline of trade, 26; revenues, 27, 52; claim on by Peruvian bondholders, 50
Guano Age, 20, 148

Habilitador, 16
Hainsworth and Company, 16
Harnecker, Otto, 36
Harvey, Robert: early career, 66-67; bank loans by, 71; requests credit from Gibbs, 72; establishes nitrate companies in London, 114
Hawkins, Bailey, 117
Hayne, James, 20, 33, 37-38
Hermann, Otto, 90
Herrera, Demofilo, 101
Herrera, Oscar: involved in Compañía Salitrera Valparaíso, 10; biography, 98-101
Humberstone, James T.: early career, 66; introduces Shanks system, 66, 69; forms nitrate companies, 68
Humberstone, J. T. y Compañía, 68

Ibañez, Adolfo, 77
Independent Liberals, 132
Industrialization, 12, 85
Inglis, George M.: early career, 66; formation of nitrate companies, 68; as manager of *oficina* Sacramento, 74; involvement in London nitrate companies, 115-16
Inquilinaje: defined, 11; as a form of labor intensive agriculture, 12; contributing to labor migration, 43; increasing demands on *inquilinos*, 80, 147-48; compared to other forced labor systems, 94
International Mercantile Bank, 67, 71, 75
Iquique: as a port, 5; effect of earthquake of 1868, 9; port facilities of, 18-20

James, Henry Berkeley, 66, 68, 115
James, Inglis and Company, 68-71
Jones, John Syers, 16

Kennedy, J.G., 136

Labor: markets for, 1, 12, 69, 148, 159; as a production cost in nitrate industry, 12; division of in nitrate industry, 13; intensity of, 12, 41, 62; productivity of, 43-44, 77, 157; in nitrate industry, 69, 121; repression of 80, 149, 152. *See also Inquilinaje*
Laisez, F., 18
Landreau, John, 52
Larraín family, 87
Liberal party, 126
Liberated Electoral, La, 135
Loans: to Peru, 20, 27; to Chile, 45
Lockett family, 72
Lopez Pérez, Arturo, 103
Lord-peasant relations, 148

Mac-Iver, David, 140, 145
Mac-Iver, Enrique: service on nitrate commission, 55; views on nitrate duty, 58-59; as critic of Balmaceda, 132; political affiliation, 141; continued service to foreign nitrate companies, 145
Marquezado, Eujenio, 10
Marsden roll-and-jaw crushers, 166n42
Martiniano Rodríguez, Luis, 138
Marx, Karl, 96, 112, 143-44, 146
Matte, Augusto, 56
Meiggs, Henry, 35
Melgar, Mescoso, 36
Merchants, Chilean, 3, 11, 84
Merchants, European, 15, 148-49
Miller, Brice Allen, 32, 134
Mining, Chilean: financing of, 12; decline of, 49; companies, 86
Mode of production, 44, 78, 150, 154
Montero Hermanos, 34-35, 117

Montt, Ambrosio, 142
Montt, Manuel, President, 4, 125-26
Montt, Pedro, President, 104
Moore, Barrington, Jr., 80
Municipal Reform Law, 144

National Agricultural Society, 80
National party, 126
Necochea, José María, 22, 98-101
Nitrate certificate holders, 55, 61, 63
Nitrate certificates: interest on, 29; acquired by Gibbs, 39; Gibbs claim on, 53; Chilean decrees on, 55; speculation in, 63-66
Nitrate companies, Chilean: organization of, 10; compared to European companies, 17-19; and *estanco*, 21-22; collapse of, 24; significance of, 148-49: Ceballos Sanz y Compañía, 75; Compañía de Salitres y Ferrocarril de Antofagasta *See* Antofagasta Nitrate and Railway Company; Compañía Salitrera America, 22, 34, 39; Compañía Salitrera California, 22, 34; Compañía Salitrera Pisagua, 14, 22, 34, 36; Compañía Salitrera Sacramento, 34, 36; Compañía Salitrera San Carlos, 10, 34, 36; Compañía Salitrera Solferino, 22, 34, 39; Compañía Salitrera Valparaíso, 10, 34; Soruco y Compañía, 14
Nitrate companies, London: 115, 119-20: Colorado Nitrate Company, 67, 114; Jazpampa Nitrate Company, 67; Lautaro Nitrate Company, 106; Liverpool Nitrate Company, 72, 114; London Nitrate Company, 115; Paccha-Jazpama Nitrate Company, 114, 122; Primitiva Nitrate Company, 114, 118; Rosario Nitrate Company, 114; San Donato Nitrate Company, 114, 116; San Jorge Nitrate Company, 114; San Pablo Nitrate Company, 114; Santa Luisa Nitrate Company, 121; Santa Rita Nitrate Company, 121; Taltal Nitrate Company, 121; Tamarugal Nitrate Company, 116, 121; Tarapacá Nitrate Company, 15; Zout Kom Nitrate Company, 116
Nitrate companies, Peruvian, 14
Nitrate consignment, 30-32
Nitrate enclave, Peruvian, 5, 8, 24
Nitrate grounds: Alianza, 65, 138-40;
Ramírez, 72; Lagunas, 74, 104
Nitrate industry production costs, 70-71
Nitrate industry productive capacity, 70
Nitrate monopoly, British, 147, 160
Nitrate of soda, 5
Nitrate policy, Chilean: factors determining 42-62; nitrate commissions, 54-55; nitrate decrees, 63-64, 66-67; effects of, 69, 75-76
Nitrate production by nationality, 73-74
Nitrate production contracts, 30-40
Nitrate Provisions and Supply Company, 117, 135
Nitrate Railways Company: organized, 34; damage to lines during War of Pacific, 64; Chilean bank accounts, 83; nullification decree, 117; acquired by North, 117; profits, 118; attack on Balmaceda, 137; breakdown of monopoly, 145
Nitrate taxes: Peruvian, 21, 27, 30, 33; Chilean, 52, 58-60, 64, 81, 128-29
Nitrate technology, 7, 12-13, 69
Nitrate trade: exports, 7, 9, 18, 64, 72-73, 113; prices, 9, 72-73, 113, 120; consumption, 18, 72-73, 113; stocks, 18, 73, 113
Nitrate workers: wages of, 69, 158-59; strikes by, 93, 121
North, John Thomas: early career, 66-67, financing of early nitrate enterprises, 71-72; formation of London companies 113-15; stock manipulations by, 118; goes to meet Balmaceda, 122; plans for nitrate trust, 140
Novoa, Jovino, 59

Oficinas, 5, 65: Agua Santa, 23, 38, 70-71, 101; Argentina, 116; Buen Retiro, 67; Carolina La, 15-16; China La, 14; Jazpampa, 67, 114; Limeña, 16, 23, 70; Noria La, 15; Paccha, 114; Palma La, 65, 116; Patria La, 70, 116; Peruana, 67; Primitiva, 67; Ramírez, 67, 114; Rosario, 116; Sacramento, 74; San Antonio, 69; San Carlos, 10; San Donato, 116; San Fernando, 75; San José de Puntunchara, 68, 71, 115; San Juan, 23, 71, 116; San Pedro, 71; Tres Marías, 71, 115
Oficinas de maquina, 7
Oficinas de parada, 7-8, 72

Olcay, R. y Compañía, 10
Oligarchy: penetration by nouveau riche, 3, and crisis of 1878, 48; attitude toward War of Pacific, 50; political power of 81; maintenance of traditional order, 82; ties to nitrate industry, 91; ties to modern and premodern economic sectors, 151. *See also* Elite
Oliva, Daniel, 8, 14, 34, 105-06
Ossa family, 84, 87

Pakenham, Frances, 51-53
Pampa del Tamarugal, 5
Paper money, 47-48
Pardo, Manuel, President: economic policies of, 20-21; and nitrate expropriation, 26; meeting with Read and Gildemeister, 31-32; and Chilean nitrate companies, 33; halts expropriation, 35
Parliamentary regime, 143-46, 152
Paunceforte, Julian, 54
Peons, 158-60
Perfetti, Pedro, 68
Periphery, 151
Peru-Bolivia Confederation, 48
Peruvian economy, 7, 24
Peruvian Nitrate Company, 65
Pinto, Aníbal, President, 45, 47, 128
Pirque, 107
Portales, Diego, 1
Prado, Mariano, General, 35, 37
Presidential system, 45, 123-25, 143, 145, 152
Procter, John, 50-54
Productivity: in nitrate industry, 13, 92, 120; in periphery, 42-43; in agriculture, 80, 92; of European capitalism, 95
Protectora, La Insurance Company, 100
Public works: Peruvian, 28; Ministry of, 81; increase in, 82, 125

Quintal, Spanish, 16

Radical party, 126
Rationalization of state functions, 45, 125
Read, Henry: views on *estanco*, 24; meeting with Pardo, 31-32; views on expropriation, 33; and formation of Tamarugal Company, 116

Rosenberg, Gustavo, 98-99, 101-02
Ross clan, 3
Ross, Gustavo, 103, 108
Ross, Jorge, 98-99, 103, 111
Rothschild, Nathan, 114

Salisbury, Lord, 50-51
Salitre, 5
Salitrera, 8
Sanfuentes, Enrique, 128
Sanguinetti, Juan, 68
Sanguinetti, J. y Compañía, 68
Santa María, Domingo, President, 81-82, 91
Santiago Tramway Company, 103
Sauce, El, 106
Schuchard and Company, 13
Shanks system, 69-70, 112
Silver prices, 86
Silver production, 86
Smith, George, 15
Social productive relations: in Chilean agriculture, 11, 43, 62, 81; in Chilean society, 93-94, 149, 159
Société Générale de Crédit Industriel et Commercial, 52, 61
South American Journal, 87, 114, 119-20
South American Steamship Company, 104
State, Chilean, 4, 124
State, Peruvian, 27
Stock exchange, London, 112
Stock exchange, Valparaíso, 10, 44, 87-89
Stock manipulation, 118-19
Subercaseaux family, 57, 84
Subercaseaux, Francisco: as founder of Compañía San Carlos, 10; family background of, 11; relationship with President Prado, 36; views on nitrate tax, 56, 59-61; relationship with Folsch and Martin, 74-75; biography, 98-99, 106-07
Sweezy, Paul, 119
Symbiotic relationship, 93, 150-59

Taltal, 34, 46-47, 56, 58-60
Tarapacá: location, 5; dependence on Chilean commerce, 10; competition with Aguas Blancas and Taltal, 56; coasting trade of, 84
Tarapacá Nitrate Company *See* T.N.C.

Tarapacá Waterworks Company, 117
T.N.C.: founding of, 15-16; negotiations with Compañía California, 22; attempted sale of, 23; accounts of, 28; profits of, 38
Toco district, 172n68

United States, 52, 61, 154
Urmeneta, Gerónimo, Senator, 49
Urrutia, Juan, 102

Valenzuela, Arturo, 126
Valparaíso, 8, 57
Valparaíso Insurance Company, 100, 103
Valparaíso Tramway Company, 98
Varela, Federico, 103
Vergara, José Francisco, 188n8
Viña Concha y Toro, 107
Vorwerk and Company, 17, 82

Wages of *inquilinos*, 80, 92
Waite, John, 72
Walker Martínez, Carlos: views on nitrate tax, 59; criticism of President Santa María, 130-31; as attorney for John Thomas North, 138; political affiliation, 141
War of the Pacific, 42-52, 63, 149
Wessel, Pedro, 139
Wheat exports, 78-79
Wheat prices, 45-46, 78
Williamson, Balfour and Company, 3, 83
Workers, skilled, 82

Zañartu;, Horacio, 138
Zavala, Santiago, 8
Zegers, Julio, 138-39, 141